Simplified
ALGEBRA I & II

David M. Kasasa

Copyright © 2013 by David M. Kasasa

ISBN-10: 1490927344
ISBN-13: 978-1490927343
LCCN: 2013912742

WHERE TO BUY THIS BOOK:
Available at Barnes and Noble Book stores
Amazon.com and all other major book stores
www.davidkasasa.com

DEDICATION

I dedicate this book to my parents who spared no penny to get me educated and without their love, commitment, dedication and sacrifice, this book could never have seen the light of day.

CONTENTS

DETAILED TABLE OF CONTENTS

Preface

I've been teaching math for over *15* years and I have
written this aid math reference book to help students
who are preparing for their placement test, Algebra *I*
and Algebra *II*.

This book provides a student with all topics in Algebra
with clear solutions, formulas and definitions.

Each topic I have worked out examples followed by many
practice problems (exercises *A*) with step by step solutions
and answers at the end of each chapter. This will prepare
students to do more practice problems (exercises *B*) on their
own which has answers at the back of this book.

Also in this book there is more than one method used
which allows a student to have different ways of
approaching each question.

You cannot succeed in any other subject in Math and
Sciences without Algebraic foundation or Algebraic skills.
This book I hope will help anyone struggling to understand
Algebra.

These are the notes I have been using to teach my Students
and I have seen great results.

David M. Kasasa
Boston MA

How to use this book:

- Avoid using a calculator unless asked for a calculator approximation.
- Take time to understand all steps in examples.
- Workout all exercises *A* before you check solutions
- Do exercises *B* and check answers at the back of the book.
- After all the above, you will be able and ready to do any question in Algebra *I* and *II*.

ACKNOWLEDGMENTS

I would like to thank the Almighty God who enabled and helped me throughout this entire production.
Secondly I would like to thank Dr. Festo Lugolobi, I could not have completed this project without his support.

CHAPTER 1

SETS AND REAL NUMBERS

1-1 Sets and Venn diagrams

A Set is a group of elements or members put into braces { }

U- Universal set is a set consisting of all the elements

A∩B- Intersection of all members of the given sets A and B

A∪B- Union of all members in both sets A and B

∈- Member of

∉- Not a member of

Ø- Empty set, null set, no member

A⊂B- A is a subset of B. all elements in set A are in B therefore,

B⊃A- B is a superset of set A.

⊄- Not a subset of

⊅- not a superset of

⊆- subset of or equal to

B'- complement of set B, all elements excluding set B

$B' \cap B = \emptyset$ $\qquad$ $B' \cup B = U$

$n(A \cup B) = n(A) + n(B) - n(A \cap B)$

<u>Venn Diagram</u>: is a diagram indicates the relationship between or among given sets.

Examples 1.1

If $A = \{1, 2, 3, 4\}$

$\quad B = \{3, 4, 5, 6\}$

$\quad C = \{7, 8, 9, 10\}$

a. Find the universal set of all the three sets A, B and C

$U = \{1, 2, 3, 4, 5, 6, 7, 8, 9, 10\}$

b. True or false
(i) $2 \in A$ __true__
(ii) $3 \in A$ and B __true__
(iii) $7 \in B$ __false__
(iv) $A \subset B$ __false__
(v) $B \subset U$ __true__
(vi) $A \not\subset U$ __false__

c. Find:
(i) $A \cap B = \{3, 4\}$
(ii) $A \cap C = \varnothing$
(iii) $A \cap B \cap C = \varnothing$
(iv) $A \cup B = \{1, 2, 3, 4, 5, 6\}$
(v) $A \cup B \cup C = U = \{1, 2, 3, 4, 5, 6, 7, 8, 9, 10\}$
(vi) $A' \cup B = \{5, 6\}$

d. Draw a Venn diagram for sets A, B and C

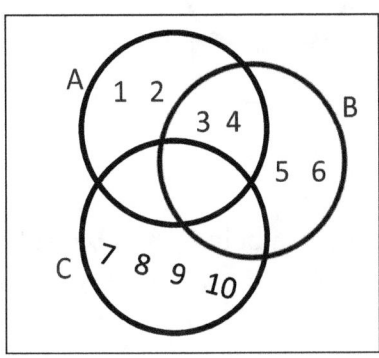

e. In a class of 25 students, 16 students like Soccer,
    10 students like both soccer and Basketball.
    Draw a Venn diagram and find how many students
    like Basketball.

Solution

Let the number of students who like Soccer be $S = 16$

Let the number of students who like Basketball be B

$$U = 25$$

$$S \cap B = 10$$

From $n(A \cup B) = n(A) + n(B) - n(A \cap B)$

$$n(S \cup B) = n(S) + n(B) - n(S \cap B)$$

But $n(S \cup B) = U = 25$

$$25 = 16 + n(B) - 10$$

$$25 = n(B) + 6$$

Subtract 6 from both sides

$$25 - 6 = n(B) + 6 - 6$$

$$19 = n(B)$$

Therefore the number of students who like Basketball is *19*

$$U = 25$$

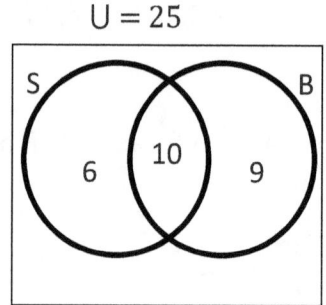

1-2 Real Numbers

Most Numbers used in Algebra belong to a big set called Real numbers. And we can find these numbers on a number line

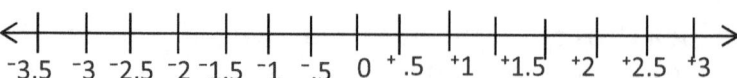

Real numbers are like a universal (U) set of all numbers and under this we have a subset called integers, also from integers we have other subsets of numbers as follows bellow:

Integers $I = \{......, -5, -4, -3, -2, -1, 0, 1, 2, 3, 4, 5 ...\}$

Natural or positive numbers $N = \{1, 2, 3, 4, 5.....\}$

Negative numbers $V = \{-1, -2, -3, -4, -5....\}$

Whole numbers $W = \{0, 1, 2, 3, 4, 5......\}$

Prime numbers $P = \{2, 3, 5....\}$ prime number is a positive integer greater than 1 and divisible by only 1 and itself.

Even numbers $E = \{...-4, -2, 0, 2, 4..\}$ a number divisible by 2.

Odd numbers $O = \{...-5, -3, -1, 1, 3, 5....\}$ is a number not divisible by 2.

Below is a Venn diagram showing all the above different sets of real numbers. Usually Venn diagrams have two or three sets, but this one has like seven sets and looks like a puzzle which will help you to know the relationship between or among the sets. All sets in a Venn diagram belong to a big set called Integers though it is not indicated in a diagram. Different sets of integers ranging from ...⁻5 to ⁺5....including real numbers like 0.5, ⅓ and √3. When you pick a number like 2, you could see that it belongs to 5 different sets of numbers.

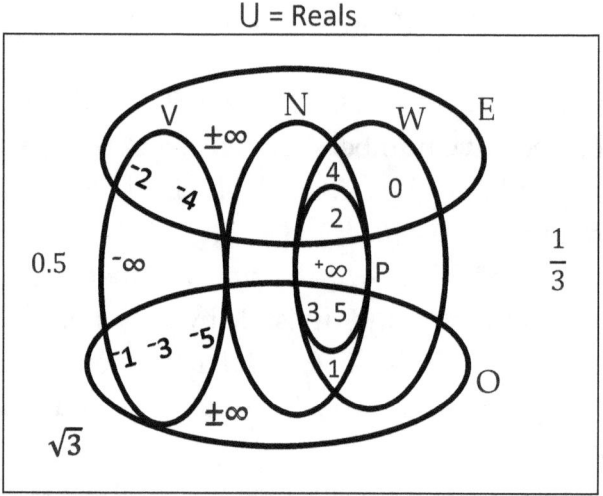

U = Reals

Figure 1.1

Examples 1.2

o From the figure *1.1*, which are the common integers or elements in set *N* and *W* but not in set *P*?

$$N \cap W \cup P' = N \cap W - P$$
$$= \{1, 2, 3, 4, 5....\} - \{2, 3, 5....\}$$
$$= \{1, 4\}$$

o Find the next three positive integers in the sequence of sets *I, N, V, W, P, E,* and set *O*.

$I = \{......,-5, -4, -3, -2, -1, 0, 1, 2, 3, 4, 5, \underline{6}, \underline{7}, \underline{8} ...\}$
$N = \{1, 2, 3, 4, 5, \underline{6}, \underline{7}, \underline{8} .....\}$
$V = \{-1, -2, -3, -4, -5, \underline{\quad}, \underline{\quad}, \underline{\quad}....\}$ no positive integers
$W = \{0, 1, 2, 3, 4, 5, \underline{6}, \underline{7}, \underline{8} ......\}$
$P = \{2, 3, 5, \underline{7}, \underline{11}, \underline{13} ....\}$

$$E = \{...-4, -2, 0, 2, 4, \underline{6}, \underline{8}, \underline{10}....\}$$
$$O = \{...-5, -3, -1, 1, 3, 5, \underline{7}, \underline{9}, \underline{11}....\}$$

1-3 Add and subtract integers using a number line

Before you could add or subtract in some other case divide or multiply any number, put in mind these two things; the signs and the word **BODMAS**, we will be using these two throughout this course.

Table 1.1   Positive and Negative signs

$+ \times + = +$	$+ \div + = +$
$- \times - = +$	$- \div - = +$
$+ \times - = -$	$+ \div - = -$
$- \times + = -$	$- \div + = -$

If a is any number then $-a = -(-a) = {}^{+}a$

$-6 - -4$ first multiply the two signs together $- \times - = +$

$-6 + 4$ it is like having or owning 4 and a debt of 6, you pay the debt and still owing a debt of 2.

Therefore ($\because$) $-6 + 4 = -2$

The word **BODMAS** stands for:

B stands for Brackets/parenthesis ( ), { }...

O stands for Of

D stands for Divide

M stands for Multiply

A stands for Add

S stands for Subtract

Use **BODMAS** to evaluate $4 - (2 \times 4) \div 4$.

Open the Brackets/parenthesis first then divide

$$4 - 8 \div 4$$
$$4 - 2 = 2$$

Example 1.3

(a) Use a number line to add or subtract the following:

(i) $5 + 4$

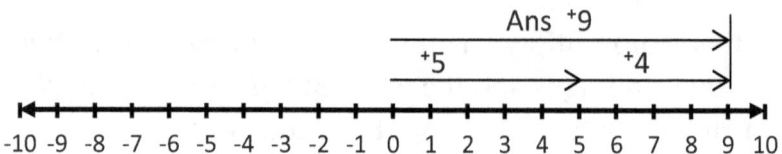

Move the arrow to the right from 0 to 5 plus 4 to right.
The answer is from zero to where the arrow stopped.

(ii) $9 + {}^-3 = 9 - 3$

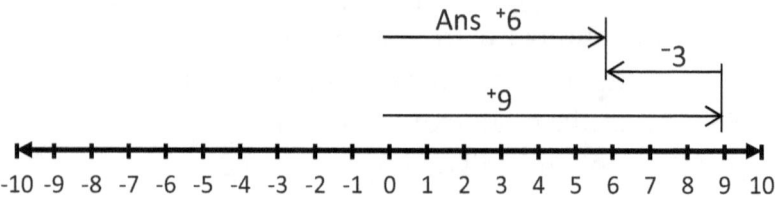

Move the arrow to the right from 0 to 9 minus 3 to left
The answer is from zero to where the arrow stopped.

(iii) $^-2 - {}^-6 = {}^-2 + 6$

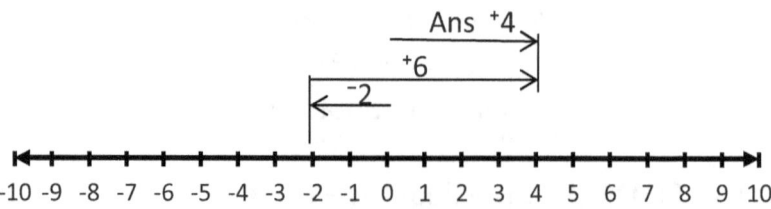

Move the arrow to the left from 0 to 2 plus 6 to right.
The answer is from zero to where the arrow stopped.

(b) Show all numbers on the number line, < 4 and > 4

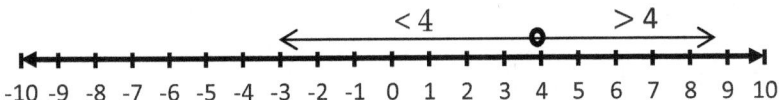

1-4 Absolute value

Absolute value of a number let it be (a) written as $|a|$ is a distance on a number line from 0 to a. it is also addictive inverse of a distance from 0 to ^-a which is a

$$^-\text{ve} \longleftarrow | \qquad | \qquad | \longrightarrow {}^+\text{ve}$$
$$\quad\ \ ^-a \qquad 0 \qquad {}^+a$$

$$|^-a| = |a| = a$$

Absolute value rules:

$|^-A| = |A| = A$

$|A|^2 = A^2$

$|AB| = |A| \times |B|$

$\left|\dfrac{A}{S}\right| = |A| \div |S|$ where $S \neq 0$

$|A| \geq 0$ where only if $A = 0$

$|AX + B| > K$

   $AX + B > K$ or $AX + B < {}^-K$

Example 1.4

(a) Find the absolute value of $^-7$

$|^-7| = |7|$
    $= 7$

(b) Find the value of the following:

(i) $\quad |2|^2 = 2^2$
$$= 2 \times 2$$
$$= 4$$

(ii) $\quad |3 \times 8| = |3| \times |8|$
$$= 3 \times 8$$
$$= 24$$

(iii) $\quad \left|\frac{12}{4}\right| = |12| \div |4|$
$$= 12 \div 4$$
$$= 3$$

(c) Simplify $|8| - 2 \times 6 \div 2$

From $|a| = a$,
$$|8| = 8$$
$$8 - 2 \times 6 \div 2$$

Using BODMAS, division first
$$8 - 2 \times 3$$
$$8 - 6$$
$$= 2$$

Exercise 1A

(1) What is the addictive inverse of $^-32$?

(2) Write a set of all prime numbers P between 20 and 40

(3) Show all numbers on number line that are >5

(4) Use a number line to evaluate $^-7 + {}^-2$

(5) Use a number line to evaluate $^-5 - {}^-11 + 2$

(6) Use a number line to evaluate $4 + (^-7) + 3$

Evaluate:

(7) $110 + (^-70) - 40$

(8) $35 \div 7 + 9 - 2 \times 5$

(9) $^-4 \times 120 \times ^-2 \times 0 \times ^-1$

(10) $-(^-2 - 7) + 11$

(11) $^-21 + 24 \div 8 \times 3$

(12) $|12| \times |10| \div 2 + ^-15$

(13) $|^-6|^2 - |4|^2$

(14) Find the integer that makes the equation a true statement.

$^-5 + 17 - 3 + \underline{} = 29$

(15) Find the next three integers in the sequence of sets p and q.

$p = \{\ldots\ldots, ^-40, ^-25, ^-10, \underline{}, \underline{}, \underline{} \ldots\}$

$q = \{\ldots 160, 80, 40, \underline{}, \underline{}, \underline{}, \ldots\}$

(16) If the sequence $X_{(n+1)} = X_n + 2$ for $n = 0, 1, 2 \ldots\ldots$

What is the value of X_3 given that $X_0 = 1$

(17) If $A = \{1, 2, 3, 4\}$

$B = \{2, 4, 6, 8\}$

$C = \{4, 5, 6, 7\}$

(a) Find the universal set of all the three sets A, B and C

(b) True or false

(i) $6 \in A$ _____

(ii) $2 \in A$ and B _____

(iii) $8 \in B$ _____

(iv) $A \subset B$ _____

(v) $B \subset U$ _____

(vi) $A \not\subset U$ _____

(c) Find:

(i) $A \cap B$

(ii) $A \cap C$

(iii) $A \cap B \cap C$

(iv) A∪B

(v) A∪B∪C

(vi) A'∪B

(d) Draw a Venn diagram for sets A, B and C

(18) A house was built from 950 feet above sea level to 1300 feet above sea level. How tall is the house?

(19) The absolute value of a number is the sum of 7 and 9 Find the number(s)

(20) From numbers 10 to 20, make a set of prime numbers P, a set of odd numbers O and a set of even numbers E. Find: (i) O∩P

(ii) P∩E

(iii) O∩P∩E

(iv) O∪P∪E

(v) O∩E∪P'

(iv) Draw a Venn diagram to illustrate sets O, P and E.

(21) In a family of 16 people, there are 10 people who like music and 3 people who like sports and music. 2 people like neither music nor sports. Use a Venn diagram to find how many people in this family like sports?

SOLUTIONS FOR EXERCISE 1A

(1) Addictive inverse of $^-32$ is $^+32$
(2)   P = {23, 29, 31, 37}
(3)   >5

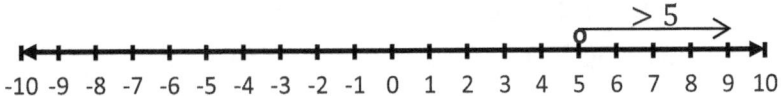

(4) $^-7 + ^-2 = ^-7 - 2$

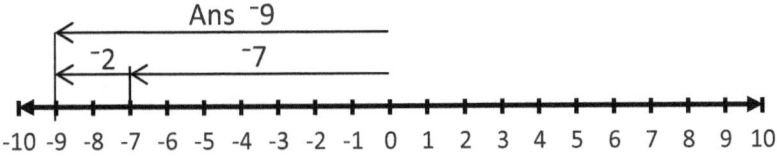

(5) $^-5 - ^-11 + 2 = ^-5 + 11 + 2$

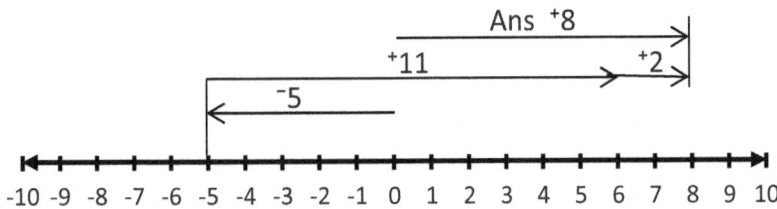

(6) $4 + (^-7) + 3 = 4 - 7 + 3$

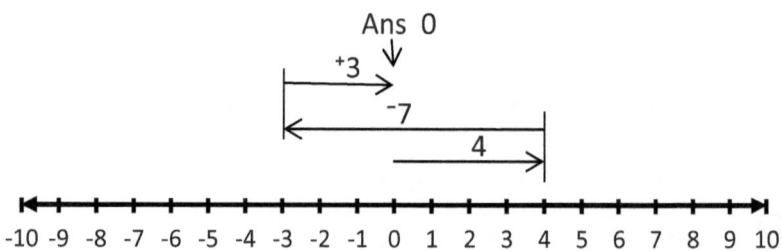

(7) $110 + (^-70) - 40$

By using **BODMAS** brackets $1^{st}(+ \times - = -)$

$110 - 70 - 40$

$110 - 110$

$= 0$

(8) $35 \div 7 + 9 - 2 \times 5$

By using **BODMAS** division comes first

$5 + 9 - 2 \times 5$

$5 + 9 - 10$

$14 - 10$

$= 4$

(9) $^-4 \times 120 \times ^-2 \times 0 \times ^-1$

Any number multiplied by *0*, the product is *0*.

$^-4 \times 120 \times ^-2 \times 0 \times ^-1$

$= 0$

(10) $-(^-2 - 7) + 11$

$-^-9 + 11$ $(- \times - = +)$

$9 + 11$

$= 20$

(11) $^-21 + 24 \div 8 \times 3$

$^-21 + 3 \times 3$

$^-21 + 9$

$= {}^-12$

(12) $|12| \times |10| \div 2 + {}^-15$

$12 \times 10 \div 2 + {}^-15$

$12 \times 5 + {}^-15$

$60 + {}^-15$

$60 - 15$

$= 45$

(13) $|{}^-6|^2 - |4|^2$

$^-6^2 - 4^2$

$({}^-6 \times {}^-6) - (4 \times 4)$

$36 - 16$

$= 20$

(14) $^-5 + 17 - 3 + \underline{} = 29$

$17 - 8 + \underline{} = 29$

$9 + \underline{20} = 29$

(15) $p = \{......, {}^-40, {}^-25, {}^-10, \underline{5}, \underline{20}, \underline{35} ...\}$

$q = \{...160, 80, 40, \underline{20}, \underline{10}, \underline{5}, ...\}$

(16) $X_0 = 1$

$X_{(n + 1)} = X_n + 2$

When $n = 0$

$X_{(0 + 1)} = X_0 + 2$

$X_1 = 1 + 2$

$X_1 = 3$

When $n = 1$

$X_{(1 + 1)} = X_1 + 2$

$X_2 = 3 + 2$

$X_2 = 5$

When $n = 2$

$X_{(2 + 1)} = X_2 + 2$

$X_3 = 5 + 2$

$X_3 = 7$

(17) If A = {1, 2, 3, 4}

B = {2, 4, 6, 8}

C = {4, 5, 6, 7}

(a) U = {1, 2, 3, 4, 5, 6, 7, 8}

b(i) 6∈A <u>false</u>

(ii) 2∈A and B <u>true</u>

(iii) 8∈B <u>true</u>

(iv) A⊂B <u>false</u>

(v) B⊂U <u>true</u>

(vi) A⊄U <u>false</u>

(c) Find:

(i) A∩B = {2, 4}

(ii) A∩C = 4

(iii) A∩B∩C = 4

(iv) A∪B = {1, 2, 3, 4, 6, 8}

(v) A∪B∪C = U = {1, 2, 3, 4, 5, 6, 7, 8}

(vi) A'∪B = {6, 8}

(d) Draw a Venn diagram for sets A, B and C

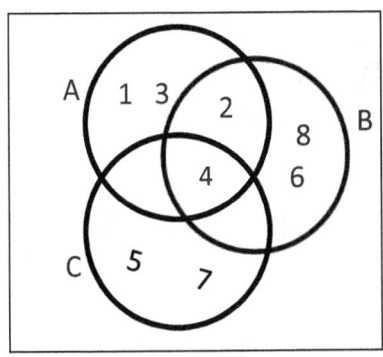

(18) 1300 – 950 = 350 feet

(19) Let the number be A
      $|A| = 7 + 9$
      $|A| = 16$ from $|{}^-A| = |A| = A$
      $A = 16$ or $A = {}^-16$
      Therefore the number(s) is ±16

(20) P = {11, 13, 17, 19}
      O = {11, 13, 15, 17, 19}
      E = {10, 12, 14, 16, 18, 20}
(i)   O∩P = {11, 13, 17, 19}
(ii) P∩E = Ø
(iii) O∩P∩E = Ø
(iv) O∪P∪E = {10, 11, 12, 13, 14, 15, 16, 17, 18, 19, 20}
(v)   O∩E∪P' = Ø

(iv) Venn diagram

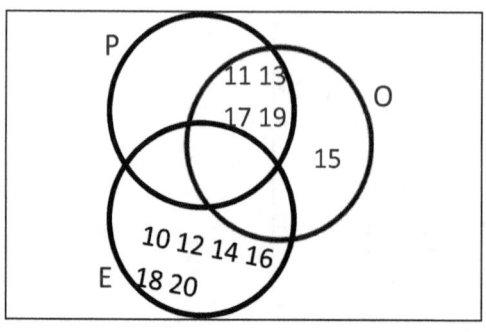

(21)   All family   U = 16
Like music M = 10
Like sports   S =?
Like music and sports
$$M \cap S = 3$$
Don't like music and no sports
$$M' \cup S' = 2$$
Like music or sports
$$M \cup S = U - M' \cup S'$$
$$M \cup S = 16 - 2 = 14$$
From $n(M \cup S) = n(S) + n(M) - n(M \cap S)$
$$14 = n(S) + 10 - 3$$
$$14 = n(S) + 7$$
Subtract 7 on both sides
$$14 - 7 = n(S) + 7 - 7$$
$$7 = n(S)$$
Therefore the number of people who like sports is 7
Venn diagram

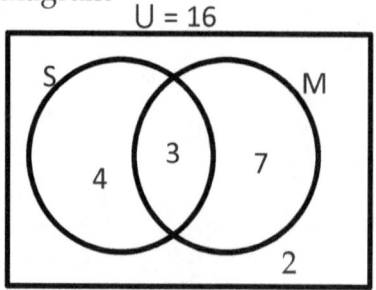

Exercise 1B

(1) What is the addictive inverse of $^-133$?

(2) Write a set of all prime numbers P between 60 and 90

(3) Show all numbers on the number line that are less than 9 and greater than 0

(4) Use a number line to evaluate $19 + {}^-15$

(5) Use a number line to evaluate $^-8 - {}^-16 + 2$

(6) Use a number line to evaluate $11 + (^-17) + 3$

Evaluate:

(7) $11[10 + (^-12 - 3)]$

(8) $^-96 \div 8 + 29 - 2 \times 7$

(9) $^-4 \times 120 \times {}^-2 \times 0 \times {}^-1$

(10) $-(^-32 - 7) + 21$

(11) $^-12 + 24 \div 6 \times 3$

(12) $(|22| \times |5|) \div 55 + {}^-15$

(13) $|9|^2 - |^-3|^2$

(14) Find the integer that makes the equation a true statement.

$^-11 + 99 - 33 + \underline{\qquad} = 29$

(15) Find the next three integers in the sequence of sets $I, N, V, W, P, E,$ and set O.

$I = \{......, ^-30, ^-25, ^-20, ^-15 \underline{\quad}, \underline{\quad}, \underline{\quad} ...\}$

$N = \{1, 2, 4, 8, \underline{\quad}, \underline{\quad}, \underline{\quad}, ...\}$

$V = \{-1, -3, -6, -10, \underline{\quad}, \underline{\quad}, \underline{\quad}, ...\}$

$W = \{0, 5, 10, 15, \underline{\quad}, \underline{\quad}, \underline{\quad}, ...\}$

$P = \{......, 5, 7, 11, 13, \underline{\quad}, \underline{\quad}, \underline{\quad}, ...\}$

$E = \{..., ^-8, ^-4, 0, 4, \underline{\quad}, \underline{\quad}, \underline{\quad}, ...\}$

$O = \{...^-7, ^-3, 1, 5, \underline{\quad}, \underline{\quad}, \underline{\quad}, ...\}$

(16) If the sequence $X_{(n + 1)} = 3X_n + 3$ for $n = 0, 1, 2, 3,......$
What is the value of X_4 given that $X_0 = 1$

(17) If $A = \{10, 12, 13, 14\}$

B = {8, 9, 11, 14}

C = {14, 15, 16, 17}

(a) Find the universal set of all the three sets A, B and C

(b) True or false

    (i) 16∈A _____

    (ii) 12∈A and B _____

    (iii) 8∈B _____

    (iv) A⊂B _____

    (v) B⊂U _____

    (vi) A⊄U _____

(c) Find:

    (i) A∩B

    (ii) A∩C

    (iii) A∩B∩C

    (iv) A∪B

    (v) A∪B∪C

    (vi) A'∪B

(d) Draw a Venn diagram for sets A, B and C

(18) A church was built from *1150* feet above sea level to *1450* feet above sea level. How tall is the church?

(19) Fill in the following integers in figure *1.1*

    {⁻9, ⁻8, ⁻7, ⁻6, 6, 7, 8, 9}

    And find N∩W∪P'

(20) The Absolute value of a number is *9* less than twice that number. Find the number

(21)

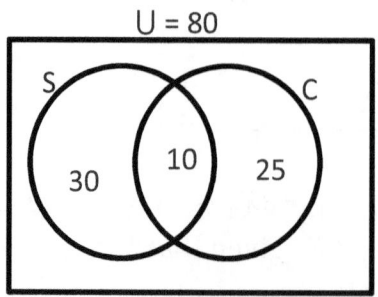

U = 80

The above Venn diagram illustrates *80* members of sports club who like to play soccer (s) and chess (c). Find: (i) members who like to play soccer (s)

      (ii) members who like to play chess (c)

      (iii) members who like to play both games

      (iv) members who do not like to play any game

(22) From numbers *20* to *35*, make a set of prime numbers *P*, a set of odd numbers *O* and a set of even numbers *E*. Find: (i) $O \cap P$

      (ii) $P \cap E$

      (iii) $O \cap P \cap E$

      (iv) $O \cup P \cup E$

      (v) $O \cap E \cup P'$

      (iv) Draw a Venn diagram to illustrate sets *O, P* and *E*.

(23) In a Camp of *21* people, there are *13* people who like music and *6* people who like sports and music. *2* people like neither music nor sports. Use a Venn diagram to find how many people in this Camp like sports.

(24) In a school of *60* students, *28* students take Science, *24* students take Math and *26* students take English class. *5* take all the three classes, *12* take Math and Science, *9* take Science and English, *7* take Math and English. Use a Venn diagram to illustrate and to find:

(i)   How many students take Math and Science only?

(ii) How many students take Math and English only?

(iii) How many students take Science and English only?

(iv) How many students take Science only?

(v) How many students take Math only?

(vi) How many students take English only?

(vii) How many students do not take any of the three Classes?

CHAPTER 2

FRACTIONS

A fraction $\dfrac{c}{d}$ where $d \neq 0$

c- Numerator

d- Denominator

$e\dfrac{c}{d}$ is a mixed fraction, which can be written as improper

fraction as $\quad \dfrac{(d \times e) + c}{d} = \dfrac{de + c}{d}$

2-1 Finding the least common denominator (LCD)

Least common denominator (LCD) is the same as the least common multiple (LCM)

We usually find LCD when adding or subtracting fractions (see examples bellow)

Given, $\dfrac{a}{b} + \dfrac{c}{d} = \dfrac{a(bd \div b) + c(bd \div d)}{bd}$

$$= \dfrac{ad + bc}{bd}$$

LCD = bd

LCD is not only got from the product of the denominators but also from the product of the prime factors of the denominators (see examples 2 bellow)

Example 2.1

$\dfrac{11}{3}, \dfrac{7}{2}$ Compare the fractions numbers and Find the bigger fraction number, also write them as mixed numbers.

Solution

$$\frac{11}{3} = \begin{array}{r} 3 \\ 3\sqrt{11} \\ \underline{-9} \\ 2 \end{array}$$

$$= 3\frac{2}{3}$$

$$\frac{7}{2} = \begin{array}{r} 3 \\ 2\sqrt{7} \\ \underline{-6} \\ 1 \end{array}$$

$$= 3\frac{1}{2}$$

The least common denominator (LCD) for 3 and 2 is 6, also the least common multiple (LCM) for 3 and 2 is 6. Multiply the two fractions with their LCD

$$\frac{11}{3} \times 6 = 22$$

$$\frac{7}{2} \times 6 = 21$$

22 is bigger than 21,

Therefore ($\therefore$) the bigger fraction is $\dfrac{11}{3}$

2-2 Adding and subtracting fractions
Examples 2.2

(a) Find the LCD and the sum of $\dfrac{1}{5}$, $\dfrac{4}{9}$, and $\dfrac{1}{3}$

Solution

3 and 5 are prime factors; we find other prime factors of 9

$$9 \atop \wedge \qquad , \; 5 \; , 3$$
$$3 \; 3$$

Multiply all prime factors of the big number and other prime factors without repeating them.

LCD = 3 × 3 × 5
LCD = 45

$$\frac{1}{5}+\frac{4}{9}+\frac{1}{3}=\frac{1(45\div5)+4(45\div9)+1(45\div3)}{45}$$

$$=\frac{9+20+15}{45}$$

$$=\frac{44}{45}$$

(b) Subtract $1\dfrac{1}{6}-\dfrac{7}{8}$

Solution

Change $1\dfrac{1}{6}$ into improper fraction

$$1\frac{1}{6}=\frac{(6\times1)+1}{6}$$

$$= \frac{6+1}{6}$$

$$= \frac{7}{6}$$

Then $\quad 1\frac{1}{6} - \frac{7}{8} = \frac{7}{6} - \frac{7}{8}$

Find the LCD of 6 and 8

Using a prime factor tree

```
        8                        6
        ∧                        ∧
      2   4                    2   3
            ∧
          2   2
```

Prime factors of $6 = 2 \times 3$

Prime factors of $8 = 2 \times 2 \times 2$

$$LCD = 2 \times 2 \times 2 \times 3$$

$$LCD = 24$$

$$\frac{7}{6} - \frac{7}{8} = \frac{7(24 \div 6) - 7(24 \div 8)}{24}$$

$$= \frac{7(4) - 7(3)}{24}$$

$$= \frac{28 - 21}{24}$$

$$= \frac{7}{24}$$

2-3 Multiplying and dividing fractions
Examples 2.3

(a) Multiply $\dfrac{8}{15} \times 1\dfrac{1}{2}$

Solution

Change $1\dfrac{1}{2}$ into improper fraction

$$1\dfrac{1}{2} = \dfrac{(2 \times 1) + 1}{2}$$

$$= \dfrac{3}{2}$$

$$\dfrac{8}{15} \times 1\dfrac{1}{2} = \dfrac{8^4}{15^5} \times \dfrac{3}{2}$$

$$= \dfrac{4}{5}$$

(b) Divide $\dfrac{4}{9} \div \dfrac{2}{3}$

When dividing fraction numbers the division sign changes into multiplication sign followed by the reciprocal of a divisor

Solution

$$\dfrac{4}{9} \div \dfrac{2}{3} = \dfrac{4^2}{9^3} \times \dfrac{3}{2}$$

$$= \dfrac{2}{3}$$

(c) Simplify $\dfrac{\dfrac{2}{3} + \dfrac{1}{4}}{\dfrac{22}{3}}$

Solution

$$\frac{\dfrac{2}{3} + \dfrac{1}{4}}{\dfrac{22}{3}} = \frac{\dfrac{8 + 3}{12}}{\dfrac{22}{3}}$$

$$= \frac{\dfrac{11}{12}}{\dfrac{22}{3}}$$

$$= \frac{11}{12} \div \frac{22}{3}$$

$$= \frac{11}{12} \times \frac{3}{22}$$

$$= \frac{1}{8}$$

(d) Given $x = 3$, $y = 4$ and $z = 12$

Find the value of $\dfrac{\dfrac{1}{x} - \dfrac{1}{y}}{\dfrac{1}{z}}$

Solution

$$\frac{\dfrac{1}{x} - \dfrac{1}{y}}{\dfrac{1}{z}} = \frac{\dfrac{1}{3} - \dfrac{1}{4}}{\dfrac{1}{12}}$$

$$= \frac{\dfrac{4-3}{12}}{\dfrac{1}{12}}$$

$$= \frac{\dfrac{1}{12}}{\dfrac{1}{12}}$$

$$= \frac{1}{12} \div \frac{1}{12}$$

$$= \frac{1}{12} \times \frac{12}{1}$$

$$= 1$$

Exercise 2A

(1) Find the bigger fraction number and the sum of:

$$\frac{^{-}2}{7} \text{ and } \frac{^{-}1}{5}.$$

(2) Find the bigger fraction number and the sum of:

$$\frac{^{-}3}{10} , \frac{^{-}5}{8} \text{ and } \frac{0}{80}$$

(3) Add $\dfrac{1}{21} + \dfrac{3}{231}$

(4) Add $\dfrac{3}{5} + \dfrac{3}{10}$

(5) Subtract $\dfrac{6}{7}$ from $1\dfrac{1}{4}$

(6) Subtract $3\dfrac{1}{4}$ from $3\dfrac{1}{2}$

(7) Subtract $\dfrac{17}{20} - \dfrac{3}{4}$

(8) Multiply $10\dfrac{4}{14} \times \dfrac{1}{12}$

(9) Divide $3\dfrac{1}{3} \div \dfrac{2}{9}$

(10) Divide $5\dfrac{1}{4} \div 3\dfrac{1}{2}$

(11) Simplify $\dfrac{\dfrac{3}{8} + \dfrac{1}{2}}{\dfrac{1}{3} + \dfrac{3}{4}}$

(12) Simplify $\dfrac{\dfrac{2}{7} + \dfrac{1}{2}}{\dfrac{2}{3} + \dfrac{1}{4}}$

(13) Simplify $\dfrac{\dfrac{3}{5} + \dfrac{3}{10}}{\dfrac{17}{20} - \dfrac{3}{4}}$

(14) Simplify $\dfrac{\dfrac{1}{5} + \dfrac{13}{60}}{\dfrac{2}{3} - \dfrac{1}{4}}$

(15) Given $x = 9$ and $y = 4$

Find the value of $\dfrac{4\left(\dfrac{y}{x} + \dfrac{1}{y}\right)}{\dfrac{5}{x}}$

(16) Given $x = 10$ and $y = 3$

SIMPLIFIED ALGEBRA I & II

Find the value of
$$\dfrac{5\left(\dfrac{y}{x} + \dfrac{1}{y}\right)}{\dfrac{5}{x}}$$

(17) Find the value of $\dfrac{3}{2x + 2y} - \dfrac{1}{x + y}$ given $(x + y) = \dfrac{1}{70}$

(18) Find the value of $\dfrac{4xy - 6x}{14y - 21}$ given $x = 3$

SOLUTIONS FOR EXERCISE 2A

(1) $\dfrac{^-2}{7}$ and $\dfrac{^-1}{5}$

7 and 5 are prime factors

LCD = 7× 5 = 35

$$\dfrac{^-2}{7} \times 35 = ^-10$$

$$\dfrac{^-1}{5} \times 35 = ^-7$$

$^-7$ is bigger than $^-10$, therefore the bigger fraction is $\dfrac{^-1}{5}$

Their sum $\quad -\dfrac{2}{7} + -\dfrac{1}{5} = -\dfrac{2}{7} - \dfrac{1}{5}$

$$= \dfrac{-2(\,35 \div 7\,) - 1(\,35 \div 5\,)}{35}$$

$$= \dfrac{-10 - 7}{35}$$

$$= \dfrac{^-17}{35}$$

(2) $\dfrac{^-3}{10}$, $\dfrac{^-5}{8}$ and $\dfrac{0}{80}$

Find the LCD of 8, 10 and 80
Using a prime factor tree

```
        80
        Λ
    2   40                      8
         Λ                      Λ                    10
       2  20                  2   4                   Λ
           Λ                      Λ                  2   5
         2  10                  2  2
             Λ
           2   5
```

Prime factors of   8 = 2 × 2 × 2
Prime factors of 10 = 2 × 5
Prime factors of 80 = 2 × 2 × 2 × 2 × 5
               LCD = 2 × 2 × 2 × 2 × 5
               LCD = 80

$$\frac{^-3}{10} \times 80 = {}^-24$$

$$\frac{^-5}{8} \times 80 = {}^-50$$

$$\frac{0}{80} \times 80 = 0$$

0 is bigger, therefore the bigger fraction is $\dfrac{0}{80}$

0 divide by any number is 0, $\dfrac{0}{80} = 0$

$$\frac{^-3}{10} + \frac{^-5}{8} + \frac{0}{80} = \frac{^-3}{10} - \frac{5}{8}$$

Prime factors of 10 = 2 × 5
Prime factors of   8 = 2 × 2 × 2

$$LCD = 2 \times 2 \times 2 \times 5$$
$$LCD = 40$$

$$\frac{^-3}{10} - \frac{5}{8} = \frac{^-3(40 \div 10) - 5(40 \div 8)}{40}$$

$$= \frac{-12 - 25}{40}$$

$$= \frac{^-37}{40}$$

(3) $\quad \dfrac{1}{21} + \dfrac{3}{231}$

Find the LCD of *21* and *231*
Using a prime factor tree

```
     231                    21
      Λ                      Λ
   3   77                  3   7
        Λ
      7  11
```

Prime factors of $\quad 21 = 3 \times 7$
Prime factors of $\quad 231 = 3 \times 7 \times 11$
$$\qquad\qquad LCD = 3 \times 7 \times 11$$
$$\qquad\qquad LCD = 231$$

$$\frac{1}{21} + \frac{3}{231} = \frac{1(231 \div 21) + 3(231 \div 231)}{231}$$

$$= \frac{1(11) + 3(1)}{24}$$

$$= \frac{11 + 3}{231}$$

$$= \frac{14}{231}$$

$$= \frac{\cancel{14}^2}{\cancel{231}^{33}}$$

$$= \frac{2}{33}$$

(4) $\dfrac{3}{5} + \dfrac{3}{10}$

Find the LCD of 5 and 10
Using a prime factor tree

$$
\begin{array}{cc}
 & 10 \\
5 & \wedge \\
 & 2 \quad 5
\end{array}
$$

5 is a Prime number
Prime factors of $\quad 10 = 2 \times 5$
$$LCD = 2 \times 5$$
$$LCD = 10$$

$$\frac{3}{5} + \frac{3}{10} = \frac{3(10 \div 5) + 3(10 \div 10)}{10}$$

$$= \frac{3(2) + 3(1)}{10}$$

$$= \frac{6 + 3}{10}$$

$$= \frac{9}{10}$$

(5) $1\frac{1}{4} - \frac{6}{7}$

$$1\frac{1}{4} - \frac{6}{7} = \frac{5}{4} - \frac{6}{7}$$

$$\text{LCD} = 4 \times 7$$
$$= 28$$

$$\frac{5}{4} - \frac{6}{7} = \frac{5(28 \div 4) - 6(28 \div 7)}{28}$$

$$= \frac{35 - 24}{28}$$

$$= \frac{11}{28}$$

(6) $3\frac{1}{2} - 3\frac{1}{4}$

$$3\frac{1}{2} - 3\frac{1}{4} = \frac{7}{2} - \frac{13}{4}$$

$$= \frac{14 - 13}{4}$$

$$= \frac{1}{4}$$

(7) $\frac{17}{20} - \frac{3}{4}$

$$\frac{17}{20} - \frac{3}{4} = \frac{17 - 15}{20}$$

$$= \frac{2}{20}$$

$$= \frac{1}{10}$$

(8) $10\frac{4}{14} \times \frac{1}{12}$

$$10\frac{4}{14} \times \frac{1}{12} = \frac{144}{14} \times \frac{1}{12}$$

$$= \frac{12}{14}$$

$$= \frac{6}{7}$$

(9) $3\frac{1}{3} \div \frac{2}{9}$

$$3\frac{1}{3} \div \frac{2}{9} = \frac{10}{3} \div \frac{2}{9}$$

Change sign followed by the reciprocal of a divisor

$$= \frac{10}{3} \times \frac{9}{2}$$

$$= 15$$

(10) $5\frac{1}{4} \div 3\frac{1}{2}$

$$5\frac{1}{4} \div 3\frac{1}{2} = \frac{21}{4} \div \frac{7}{2}$$

$$= \frac{\cancel{21}^3}{\cancel{4}^2} \times \frac{\cancel{2}}{\cancel{7}}$$

$$= \frac{3}{2}$$

$$= 1\frac{1}{2}$$

(11) $$\frac{\dfrac{3}{8} + \dfrac{1}{2}}{\dfrac{1}{3} + \dfrac{3}{4}}$$

LCD of 8 and 2 is 8 and also LCD of 3 and 4 is 12

$$\frac{\dfrac{3}{8} + \dfrac{1}{2}}{\dfrac{1}{3} + \dfrac{3}{4}} = \frac{\dfrac{3(8 \div 8) + 1(8 \div 2)}{8}}{\dfrac{1(12 \div 3) + 3(12 \div 4)}{12}}$$

$$= \frac{\dfrac{3 + 4}{8}}{\dfrac{4 + 9}{12}}$$

$$= \frac{\dfrac{7}{8}}{\dfrac{13}{12}}$$

$$= \frac{7}{8} \div \frac{13}{12}$$

$$= \frac{7}{8} \times \frac{12}{13}$$

$$= \frac{21}{26}$$

(12) $\dfrac{\dfrac{2}{7} + \dfrac{1}{2}}{\dfrac{2}{3} + \dfrac{1}{4}}$

$$\frac{\dfrac{2}{7} + \dfrac{1}{2}}{\dfrac{2}{3} + \dfrac{1}{4}} = \frac{\dfrac{2(14 \div 7) + 1(14 \div 2)}{14}}{\dfrac{2(12 \div 3) + 1(12 \div 4)}{12}}$$

LCD of 7 and 2 is 14 and also LCD of 3 and 4 is 12

$$= \frac{\dfrac{4 + 7}{14}}{\dfrac{8 + 3}{12}}$$

$$= \frac{\dfrac{11}{14}}{\dfrac{11}{12}}$$

$$= \frac{11}{14} \div \frac{11}{12}$$

$$= \frac{11}{14} \times \frac{12}{11}$$

$$= \frac{6}{7}$$

(13)

$$\frac{\dfrac{3}{5} + \dfrac{3}{10}}{\dfrac{17}{20} - \dfrac{3}{4}}$$

LCD of *5* and *10* is *10* and also LCD of *20* and *4* is *20*

$$\frac{\dfrac{3}{5} + \dfrac{3}{10}}{\dfrac{17}{20} - \dfrac{3}{4}} = \frac{\dfrac{3(10 \div 5) + 3(10 \div 10)}{10}}{\dfrac{17(20 \div 20) - 3(20 \div 4)}{20}}$$

$$= \frac{\dfrac{6 + 3}{10}}{\dfrac{17 - 15}{20}}$$

$$= \frac{\dfrac{9}{10}}{\dfrac{2}{20}}$$

$$= \frac{9}{10} \div \frac{2}{20}$$

$$= \frac{9}{10} \times \frac{20}{2} = 9$$

(14) $$\dfrac{\dfrac{1}{5} + \dfrac{13}{60}}{\dfrac{2}{3} - \dfrac{1}{4}}$$

LCD of 5 and 60 is 60 and also LCD of 3 and 4 is 12

$$\dfrac{\dfrac{1}{5} + \dfrac{13}{60}}{\dfrac{2}{3} - \dfrac{1}{4}} = \dfrac{\dfrac{1(60 \div 5) + 13(60 \div 60)}{60}}{\dfrac{2(12 \div 3) - 1(12 \div 4)}{12}}$$

$$= \dfrac{\dfrac{12 + 13}{60}}{\dfrac{8 - 3}{12}}$$

$$= \dfrac{\dfrac{25}{60}}{\dfrac{5}{12}}$$

$$= \dfrac{25}{60} \div \dfrac{5}{12}$$

$$= \dfrac{25^{5}}{60^{12}} \times \dfrac{12}{5}$$

$$= 1$$

(15) $$\frac{4\left(\frac{y}{x}+\frac{1}{y}\right)}{\frac{5}{x}}=\frac{4\left(\frac{4}{9}+\frac{1}{4}\right)}{\frac{5}{9}}$$

LCD of 9 and 4 is 36

$$=\frac{4\left(\frac{4(36\div9)+1(36\div4)}{36}\right)}{\frac{5}{9}}$$

$$=\frac{4\left(\frac{16+9}{36}\right)}{\frac{5}{9}}$$

$$=\frac{4\left(\frac{25}{36}\right)}{\frac{5}{9}}$$

$$=\frac{\frac{25}{9}}{\frac{5}{9}}$$

$$=\frac{25}{9}\div\frac{5}{9}$$

$$=\frac{25}{9}\times\frac{9}{5}$$

$$=5$$

(16)
$$\frac{5\left(\frac{y}{x} + \frac{1}{y}\right)}{\frac{5}{x}} = \frac{5\left(\frac{3}{10} + \frac{1}{3}\right)}{\frac{5}{10}}$$

LCD of *10* and *3* is *30*

$$= \frac{5\left(\frac{3(30 \div 10) + 1(30 \div 3)}{30}\right)}{\frac{5}{10}}$$

$$= \frac{5\left(\frac{9 + 10}{30}\right)}{\frac{1}{2}}$$

$$= \frac{5\left(\frac{19}{30}\right)}{\frac{1}{2}}$$

$$= \frac{\frac{19}{6}}{\frac{1}{2}}$$

$$= \frac{19}{6} \div \frac{1}{2}$$

$$= \frac{19}{6} \times \frac{2}{1}$$

$$= \frac{19}{3}$$

(17) $\quad \dfrac{3}{2x + 2y} - \dfrac{1}{x + y} = \dfrac{3}{2(x + y)} - \dfrac{1}{x + y}$

$$= \frac{3 - 2}{2(x + y)}$$

$$= \frac{1}{2(x + y)}$$

But $(x + y) = \dfrac{1}{70}$

$$\frac{1}{2(x + y)} = \frac{1}{2 * \frac{1}{70}}$$

$$= \frac{1}{\frac{1}{35}}$$

$$= 1 \div \frac{1}{35}$$

$$= 1 \times \frac{35}{1}$$

$$= 35$$

(18) $\quad \dfrac{4xy - 6x}{14y - 21} = \dfrac{2x(2y - 3)}{7(2y - 3)}$

$$= \frac{2x}{7}$$

$$= \frac{2 \times 3}{7}$$

$$= \frac{6}{7}$$

Exercise 2B

(1) Find the bigger fraction number and the sum of:

$\dfrac{^-7}{8}$ and $\dfrac{^-7}{6}$.

(2) Add $\dfrac{1}{4} + \dfrac{3}{22}$

(3) Add $\dfrac{1}{5} + \dfrac{3}{4} + \dfrac{2}{3}$

(4) Subtract $\dfrac{^-4}{5}$ from $1\dfrac{1}{3}$

(5) Subtract $2\dfrac{1}{2}$ from $^-3\dfrac{1}{3}$

(6) Subtract $1\dfrac{1}{12} - \dfrac{3}{4} - \dfrac{^-1}{3}$

(7) Multiply $10\dfrac{1}{12} \times \dfrac{6}{11}$

(8) Divide $\quad 4\dfrac{1}{5} \div \dfrac{7}{15}$

(9) Divide $\quad 3\dfrac{1}{3} \div 5\dfrac{1}{2} \div \dfrac{10}{11}$

(10) Simplify $\quad \dfrac{\dfrac{3}{16} + \dfrac{1}{2}}{3\dfrac{7}{16}}$

(11) Simplify $\quad \dfrac{\dfrac{7}{16} + \dfrac{1}{2}}{\dfrac{3}{8} + \dfrac{1}{4}}$

(12) Simplify $\quad \dfrac{\dfrac{5}{6} + \dfrac{3}{8}}{1\dfrac{11}{24} - \dfrac{1}{4}}$

(13) Simplify $\quad \dfrac{\dfrac{5}{6} - \dfrac{11}{60}}{\dfrac{1}{4} - \dfrac{11}{12}}$

(14) Given $\quad x = {}^-2$ and $y = {}^-11$

Find the value of $\quad \dfrac{\dfrac{x}{y} + \dfrac{1}{xy}}{\dfrac{1}{2y}}$

(15) Given $x = 3$ and $y = 2$

Find the value of $\dfrac{6\left(\dfrac{y}{x} + \dfrac{1}{y}\right)}{\dfrac{7}{2x}}$

(16) Find the value of $\dfrac{2}{x+y} - \dfrac{1}{3x+3y}$ given $(x + y) = \dfrac{5}{12}$

(17) Find the value of $\dfrac{^-10xy + 5x}{12y - 6}$ given $x = 1$

(18) Simplify $\dfrac{1}{2x} + \dfrac{1}{y} - \dfrac{1}{4x} - \dfrac{3}{4y}$

(19) Simplify $\dfrac{ab}{d} + \dfrac{ab}{c}$

(20) Simplify $\dfrac{abc}{de} \div \dfrac{ba}{ed}$

CHAPTER 3

EXPONENTS OR POWER

If $a^m = n$, therefore, a is the base and m is the exponent or power that shows the number of times the base is to be multiplied by itself and n is the result or the product.

3-1 Laws of Exponents
Examples 3.1
Simplify:

(a) 777^0

From the law of exponents $a^0 = 1$ any number to power zero is 1

$$777^0 = 1$$

(b) $3^2 \times 3^3$

From the law of exponents $a^m \times a^n = a^{m+n}$

$$3^2 \times 3^3 = 3^{2+3}$$
$$= 3^5$$
$$= 3 \times 3 \times 3 \times 3 \times 3$$
$$= 243$$

(c) $\dfrac{5^4}{5}$

From the law of exponents $\dfrac{a^m}{a^n} = a^m \div a^n = a^{m-n}$

$$\frac{5^4}{5} = 5^4 \div 5 = 5^{4-1}$$
$$= 5^3$$

$$= 5 \times 5 \times 5$$
$$= 125$$

(d) $(2^2)^3$

From the law of exponents $(a^m)^n = a^{mn}$

$$(2^2)^3 = 2^{2 \times 3}$$
$$= 2^6$$
$$= 64$$

(e) $(2 \times 5)^2$

From the law of exponents $(ab)^m = a^m b^m$

$$(2 \times 5)^2 = 2^2 \times 5^2$$
$$= 4 \times 25$$
$$= 100$$

(f) $\left(\dfrac{2}{3}\right)^2$

From the law of exponents $\left(\dfrac{a}{b}\right)^m = \dfrac{a^m}{b^m} \qquad b \neq 0$

$$\left(\dfrac{2}{3}\right)^2 = \dfrac{2^2}{3^2}$$

$$= \dfrac{4}{9}$$

(g) 4^{-2}

From the law of exponents $a^{-n} = \dfrac{1}{a^n}$

$$4^{-2} = \dfrac{1}{4^2}$$

$$= \dfrac{1}{16}$$

(h) $\left(\dfrac{4}{3}\right)^{-2}$

From the law of exponents $\left(\dfrac{a}{b}\right)^{-n} = \left(\dfrac{b}{a}\right)^{n}$

$$\left(\dfrac{4}{3}\right)^{-2} = \left(\dfrac{3}{4}\right)^{2}$$

$$= \dfrac{3^2}{4^2}$$

$$= \dfrac{9}{16}$$

(i) $\dfrac{3^{-2}}{2^{-3}}$

From the law of the exponents $\dfrac{a^{-n}}{b^{-m}} = \dfrac{b^m}{a^n}$

$$\dfrac{3^{-2}}{2^{-3}} = \dfrac{2^3}{3^2}$$

$$= \dfrac{8}{9}$$

(j) $3^{\frac{4}{2}}$

From the law of exponents $a^{\frac{m}{n}} = \sqrt[n]{a^m}$

$$3^{\frac{4}{2}} = \sqrt[2]{3^4}$$

$$= \sqrt{3 \times 3 \times 3 \times 3}$$

$$= \sqrt{81}$$

$$= 9$$

(k) $(8p^3q^9)^{\frac{1}{3}} = (2^3p^3q^9)^{\frac{1}{3}}$

$$= (2pq^3)^{\frac{1}{3} \times 3}$$

$$= 2pq^3$$

3-2 Scientific notation

Scientific notation is written in this form $a \times 10^n$ where $1 \leq |a| < 10$ and n is an integer.

Below is the table showing that when a decimal point moves to the right ($\rightarrow$), the Power is Negative and when it moves to the left ($\leftarrow$), the Power is Positive.

This table helps in studying and understanding the concept in scientific notation, exponents or powers and decimals (see next chapter for Decimals).

Table 3.1 _____

$$0.0001 = 0.0001 = 1.0 \times 10^{-4} = 10^{-4} = \frac{1}{10^4} = \frac{1}{10000}$$

$$0.001 = 0.001 = 1.0 \times 10^{-3} = 10^{-3} = \frac{1}{10^3} = \frac{1}{1000}$$

$$0.01 = 0.01 = 1.0 \times 10^{-2} = 10^{-2} = \frac{1}{10^2} = \frac{1}{100}$$

$$0.1 = 0.1 = 1.0 \times 10^{-1} = 10^{-1} = \frac{1}{10^1} = \frac{1}{10}$$

$$1 = 1.0 = 1.0 \times 10^0 = 10^0$$

$$10 = 10.0 = 1.0 \times 10^1 = 10^1$$

$$100 = 100.0 = 1.0 \times 10^2 = 10^2$$

$$1000 = 1000.0 = 1.0 \times 10^3 = 10^3$$

$$10000 = 10000.0 = 1.0 \times 10^4 = 10^4$$

Examples 3.2

(a) Write *200* in scientific notation.

Solution

200 = 200.0

Move decimal point to the left two places. Power is $^+2$ because the point moved two places to the left

$$200.0 = 2.0 \times 10^2$$
$$= 2 \times 10^2$$

(b) Write *0.0014* in scientific notation.

Solution

Move decimal point to the right hand side three places
The Power is $^-3$ because the point moved 3 places to the right

$$0.0014 = 1.4 \times 10^{-3}$$

Exercise 3A

Simplify the following numbers:

(1) 2013^0
(2) $7^3 \times 7^{-1} \times 7^0$
(3) $8^5 \div 8^2$
(4) $(5^2)^2$

(5) $(2x)^3$

(6) $x^6y * x^2y^5$

(7) $9m^{10}n \div 3m^2n^{-2}$

(8) $\left(\dfrac{3}{4}\right)^2$

(9) 64×2^{-3}

(10) $\left(\dfrac{9}{5}\right)^{-2}$

(11) $\dfrac{2^{-3}}{4^{-2}}$

(12) $6^{\frac{4}{2}}$

(13) $7(2p^2q^4 + 2p^2q^4)^{\frac{1}{2}}$

(14) $4^{-\pi} * 2^{2+2\pi}$

(15) $2^{\frac{7}{4}} - 2^{\frac{3}{4}}$

Write the following in scientific notation:

(16) 2013

(17) 0042.3

(18) 0.0062

(19) 88×100

(20) 0.0005760

SOLUTIONS FOR EXERCISE 3A

(1) $2013^0 = 1$ any number to power zero is one

(2) $7^3 \times 7^{-1} \times 7^0$
$$= 7^{(3 + {}^-1 + 0)}$$
$$= 7^2$$
$$= 7 \times 7$$
$$= 49$$

(3) $8^5 \div 8^2$
$$= 8^{(5 - 2)}$$
$$= 8^3$$
$$= 8 \times 8 \times 8$$
$$= 512$$

(4) $(5^2)^2$
$$= 5^{(2 \times 2)}$$
$$= 5^4$$
$$= 5 \times 5 \times 5 \times 5$$
$$= 625$$

(5) $(2x)^3 = 2^3 x^3$
$$= 8x^3$$

(6) $x^6 y * x^2 y^5$
$$= x^{(6 + 2)} y^{(1 + 5)}$$
$$= x^8 y^6$$

(7) $9m^{10} n \div 3m^2 n^{-2}$
$$= 3^2 m^{10} n \div 3m^2 n^{-2}$$

$$= 3^{(2-1)}m^{(10-2)}n^{(1-{}^-2)}$$
$$= 3^{(2-1)}m^{(10-2)}n^{(1+2)}$$
$$= 3m^8n^3$$

(8) $\left(\dfrac{3}{4}\right)^2 = \dfrac{3^2}{4^2}$

$$= \dfrac{9}{16}$$

(9) 64×2^{-3}

$$= 2^6 \times 2^{-3}$$
$$= 2^{(6+{}^-3)}$$
$$= 2^{(6-3)}$$
$$= 2^3$$
$$= 8$$

(10) $\left(\dfrac{9}{5}\right)^{{}^-2} = \dfrac{9^{-2}}{5^{-2}}$

$$= \dfrac{5^2}{9^2}$$

$$= \dfrac{25}{81}$$

(11) $\dfrac{2^{-3}}{4^{-2}}$

From
$$\dfrac{a^{-n}}{b^{-m}} = \dfrac{b^m}{a^n}$$

$$\dfrac{2^{-3}}{4^{-2}} = \dfrac{4^2}{2^3}$$

$$= \dfrac{16}{8} = 2$$

(12) $6^{\frac{4}{2}}$

    From

$$a^{\frac{m}{n}} = \sqrt[n]{a^m}$$

$$6^{\frac{4}{2}} = \sqrt[2]{6^4}$$

$$= \sqrt{6 \times 6 \times 6 \times 6}$$

$$= \sqrt{1296}$$

$$= 36$$

(13) $7(2p^2q^4 + 2p^2q^4)^{\frac{1}{2}}$

$$= 7(4p^2q^4)^{\frac{1}{2}}$$

$$= 7(2^2p^2q^4)^{\frac{1}{2}}$$

$$= 7(2\,pq^2)^{\frac{1}{2} \times 2}$$

$$= 14pq^2$$

(14) $4^{-\pi} * 2^{2+2\pi}$

$$= (2^2)^{-\pi} * 2^{2+2\pi}$$

$$= 2^{-2\pi} * 2^{2+2\pi}$$

$$= 2^{-2\pi + 2 + 2\pi}$$

$$= 2^2$$

$$= 4$$

(15) $2^{\frac{7}{4}} - 2^{\frac{3}{4}}$

$$= \left(2^{\frac{1}{4}}\right)^7 - \left(2^{\frac{1}{4}}\right)^3$$

$$= \left(2^{\frac{1}{4}}\right)^3 \left(2^{\frac{1}{4}}\right)^4 - \left(2^{\frac{1}{4}}\right)^3$$

$$= \left(2^{\frac{1}{4}}\right)^3 \left[\left(2^{\frac{1}{4}}\right)^4 - 1\right]$$

$$= \left(2^{\frac{1}{4}}\right)^{3} (2 - 1)$$

$$= \left(2^{\frac{1}{4}}\right)^{3}$$

$$= 2^{\frac{3}{4}}$$

(16) $2013 = 2013.0$

$2013.0 = 2.013 \times 10^{3}$

> The Power is ⁺3 because the point moved 3 places to the left

(17) $0042.3 = 42.3$

$42.3 = 4.23 \times 10^{1}$

> The power is ⁺1 because the point moved one place to the left

(18) $0.0062 = 6.2 \times 10^{-3}$

> The power is ⁻3 because the point moved 3 places to the right

(19) $88 \times 100 = 88.0 \times 10^{2}$

$88.0 \times 10^{2} = 8.8 \times 10^{1} \times 10^{2}$
$$= 8.8 \times 10^{1 +2}$$
$$= 8.8 \times 10^{3}$$

(20) $0.0005760 = 5.76 \times 10^{-4}$

> The power is ⁻4 because the point moved 4 places to the right

Exercise 3B

Simplify the following numbers:

(1) 1^{11}
(2) $10^4 \times 10^3 \div 10^5$
(3) $4^2 \div 4^{-2}$
(4) $(2^3)^3$
(5) $(3xy)^2$
(6) $x^{-4}y * x^2y$
(7) $4p^4q \div 2p^2q^{-2}$

(8) $\left(\dfrac{2}{3}\right)^2$

(9) 81×3^{-3}

(10) $\left(\dfrac{1}{6}\right)^{-2}$

(11) $\dfrac{3^{-3}}{9^{-2}}$

(12) $25^{\frac{3}{2}}$

(13) $3(8p^4q^2 + 8p^4q^2)^{\frac{1}{2}}$

(14) $8^{-\pi} * 2^{3(1+\pi)}$

(15) $4^{\frac{4}{3}} - 2^{\frac{5}{3}}$

Write the following in scientific notation:

SIMPLIFIED ALGEBRA I & II

(16) 1213
(17) 0.576
(18) 0.00082
(19) 543.18
(20) 099990
(21) 10

CHAPTER 4

DECIMALS

From a fraction we can have a decimal, for example

$$\frac{1}{10} = \begin{array}{r} 0.1 \\ 10\sqrt{10} \\ -10 \\ \hline - \ - \end{array} \qquad \frac{1}{10} = 0.1$$

0.1 is the example of a **Terminating Decimal**

$$\frac{1}{3} = \begin{array}{r} 0.33\ .... \\ 3\sqrt{10} \\ -9 \\ \hline 10 \\ -9 \\ \hline 1 \end{array} \qquad \frac{1}{3} = 0.33....$$

0.33… is the example of a **Repeating Decimal**

Decimals which are neither Terminating non Repeating are called **Irrational Numbers**

For example $\frac{1}{13} = 0.076923.....$

4-1 Convert a Fraction into a Decimal Numbers
Examples 4.1

(a) Convert these four fractions, $\frac{1}{8}, \frac{1}{9}, \frac{1}{11},$ and $\frac{1}{19}$ into decimals and also determine whether it is a terminating, irrational or repeating decimal.

Solution

$$\frac{1}{8} = \begin{array}{r} 0.125 \\ 8\sqrt{10} \\ -8 \\ \hline 20 \\ -16 \\ \hline 40 \\ -40 \\ \hline -- \end{array}$$

$$\frac{1}{9} = \begin{array}{r} 0.11\ldots \\ 9\sqrt{10} \\ -9 \\ \hline 10 \\ -9 \\ \hline 1 \end{array}$$

$$\frac{1}{11} = \begin{array}{r} 0.0909\ldots \\ 11\sqrt{100} \\ -99 \\ \hline 100 \\ -99 \\ \hline 1 \end{array}$$

$$\frac{1}{19} = \begin{array}{r} 0.0526\ldots \\ 19\sqrt{100} \\ -95 \\ \hline 50 \\ -38 \\ \hline 120 \\ -114 \\ \hline 6 \end{array}$$

$\dfrac{1}{8} = 0.125$ it is Terminating Decimal

$\dfrac{1}{9} = 0.11\ldots\ldots$it is Repeating Decimal

$\dfrac{1}{11} = 0.0909\ldots$it is Repeating Decimal

$\dfrac{1}{19} = 0.0526\ldots$it is Irrational Number

(b) Given two decimal numbers 0.154 and 0.159

(i) What is the bigger number?

    0.159 is bigger than 0.154

(ii) Round off to hundredth or 2 decimal places

      $0.154 \approx 0.15$

      $0.159 \approx 0.16$

4-2 Adding and Subtracting Decimal Numbers

When Adding or Subtracting a Decimal number, write all the numbers in a column having decimal points in an up straight line.

Examples 4.2

(a) $71.21 + 9.8$

$$\begin{array}{r} 71.21 \\ +9.80 \\ \hline 81.01 \end{array}$$

(b) $87.66 - 53.208$

$$\begin{array}{r} 87.66 \\ -53.208 \\ \hline 34.452 \end{array}$$

4-3 Multiplying and Dividing Decimal Numbers

Examples 4.3

(a) $6.28 \times 5.5 =$

$$\begin{array}{r} 6.28 \quad (\textit{2 decimal places}) \\ \times 5.5 \quad (\textit{1 decimal place}) \\ \hline 3140 \\ 3140 \\ \hline 34.540 \quad (\textit{3 decimal places}) \end{array}$$

$6.28 \times 5.5 = 34.54$

OR this can be solved by removing the decimal points first and can be done by using the ideas learned in table *3.1* in chapter *3*.

$$6.28 = 628 \times 10^{-2}$$
$$5.5 = 55 \times 10^{-1}$$
$$6.28 \times 5.5 = 628 \times 10^{-2} \times 55 \times 10^{-1}$$
$$= 628 \times 55 \times 10^{-2} \times 10^{-1}$$

$$= 34540 \times 10^{-3}$$
$$= 34.540$$
$$= 34.54$$

(b) $28.416 \div 3.20$

move the decimal point one place to the right of a divisor and also a dividend

$$28.416 \div 3.20 = 284.16 \div 32$$

$$
\begin{array}{r}
8.88 \\
32\overline{)284.16} \\
-256 \\
\hline
281 \\
-256 \\
\hline
256 \\
-256 \\
\hline
- - - \\
\end{array}
$$

OR this can be solved by removing the decimal points first and can be done by using the ideas learned in table *3.1* in chapter *3*.

$$28.416 \div 3.20 = \frac{28.416}{3.20}$$

$$\frac{28.416}{3.20} = \frac{28416 \times 10^{-3}}{32 \times 10^{-1}}$$

$$= \frac{28416}{32} \times \frac{10^{-3}}{10^{-1}}$$

$$= 888 \times 10^{(-3 - ^-1)}$$
$$= 888 \times 10^{(-3 + 1)}$$

$$= 888 \times 10^{-2}$$
$$= 8.88$$

(c) 0.03×0.66

$$
\begin{array}{rl}
0.03 & (\textit{2 decimal places}) \\
\times\, 0.66 & (\textit{2 decimal place}) \\
\hline
018 & \\
018 & \\
000 & \\
\hline
0.0198 & (\textit{4 decimal places})
\end{array}
$$

$$0.03 \times 0.66 = 0.0198$$

OR this can be solved by removing the decimal points first and can be done by using the ideas learned in table *3.1* in chapter *3*.

$$0.03 = 3 \times 10^{-2}$$
$$0.66 = 66 \times 10^{-2}$$
$$0.03 \times 0.66 = 3 \times 10^{-2} \times 66 \times 10^{-2}$$
$$= 3 \times 66 \times 10^{-2} \times 10^{-2}$$
$$= 198 \times 10^{-4}$$
$$= 0.0198$$

(d) $0.66 \div 0.03$

move the decimal point one place to the right of a divisor and also a dividend

$$0.66 \div 0.03 = 66 \div 3$$

$$\begin{array}{r} 22 \\ 3\sqrt{66} \\ -6 \\ \hline 6 \\ -6 \\ \hline - \end{array}$$

OR this can be solved by removing the decimal points first and can be done by using the ideas learned in table *3.1* in chapter *3*.

$$0.66 \div 0.03 = \frac{0.66}{0.03}$$

$$\frac{0.66}{0.03} = \frac{66 \times 10^{-2}}{3 \times 10^{-2}}$$

$$= \frac{66}{3} \times \frac{10^{-2}}{10^{-2}}$$

$$= 22 \times 10^{(-2 - {}^{-}2)}$$
$$= 22 \times 10^{(-2 + 2)}$$
$$= 22 \times 10^{0}$$
$$= 22 \times 1$$
$$= 22$$

Exercise 4A

(1) Convert the following fractions into decimals and determine whether it is a Terminating Decimal, Repeating Decimal or Irrational number (round off your Answers to the nearest a thousandth or three decimal places).

(a) $\dfrac{1}{31}$

(b) $\dfrac{1}{16}$

(c) $\dfrac{3}{11}$

(2) Given two decimal numbers 0.154 and 0.0154
(i) What is the bigger number?
(ii) Round off the two decimal numbers to hundredth or 2 decimal places.
(3) What is the bigger decimal number -0.0721 and -0.0731
(4) Add or subtract the following Decimals:
(a) $0.989 + 0.011$
(b) $4.278 + 32.2375 + 3.6845$
(c) $25.62 - 9.42$
(d) $12 - 0.02$
(5) Multiply or divide the following Decimals:
(a) 4.2×7.52
(b) 5.4×0.6
(c) $0.54 \div 0.06$
(d) $12.126 \div 25.8$
(e) $99 \div 0.11$

SOLUTIONS FOR EXERCISE 4A

1(a) $\dfrac{1}{31} =$

$$
\begin{array}{r}
0.032 \\
31\overline{\smash{\big)}100} \\
-93 \\
\hline
70 \\
-62 \\
\hline
8
\end{array}
$$

$\dfrac{1}{31} = 0.032$ It is Irrational number

(b) $\dfrac{1}{16} =$

$$
\begin{array}{r}
0.0625 \\
16\overline{\smash{\big)}100} \\
-96 \\
\hline
40 \\
-32 \\
\hline
80 \\
-80 \\
\hline
- \; -
\end{array}
$$

$\dfrac{1}{16} = 0.063$ It is Terminating Decimal

(c) $\dfrac{3}{11} =$

$$
\begin{array}{r}
0.2727 \ldots \\
11\overline{\smash{\big)}30} \\
-22 \\
\hline
80 \\
-77 \\
\hline
30 \\
-22 \\
\hline
80 \\
-77 \\
\hline
3
\end{array}
$$

$\dfrac{3}{11} = 0.273$ It is Repeating Decimal

2(i)   *0.154* is bigger than *0.0154*

(ii) $0.154 \approx 0.15$

$0.0154 \approx 0.02$

(3) The bigger decimal number between ⁻*0.0721* and ⁻*0.0731* is ⁻*0.0721*.

4(a) 0.989 + 0.011

$$\begin{array}{r} 0.989 \\ +0.011 \\ \hline 1.000 \end{array}$$

$0.989 + 0.011 = 1$

(b) 4.278 + 32.2375 + 3.6845

$$\begin{array}{r} 4.2780 \\ 32.2375 \\ +\ 3.6845 \\ \hline 40.2000 \end{array}$$

$4.278 + 32.2375 + 3.6845 = 40.2$

(c)   25.62 – 9.42

$$\begin{array}{r} 25.62 \\ -9.42 \\ \hline 16.2 \end{array}$$

$25.62 - 9.42 = 16.2$

(d)   12 – 0.02

$$\begin{array}{r} 12.00 \\ -0.02 \\ \hline 11.98 \end{array}$$

$12 - 0.02 = 11.98$

5(a) 4.2×7.5

$$
\begin{array}{r}
4.2 \\
\times\ 7.52 \\
\hline
84 \\
210 \\
294 \\
\hline
31.584
\end{array}
$$

4.2 (1 *decimal places*)
$\times\ 7.52$ (2 *decimal place*)

31.584 (3 *decimal places*)

$4.2 \times 7.52 = 31.584$

OR this can be solved by removing the decimal points first and can be done by using the ideas learned in table *3.1* in chapter *3*.

$$4.2 = 42 \times 10^{-1}$$
$$7.52 = 752 \times 10^{-2}$$
$$4.2 \times 7.52 = 42 \times 10^{-1} \times 752 \times 10^{-2}$$
$$= 42 \times 752 \times 10^{-1} \times 10^{-2}$$
$$= 31584 \times 10^{-3}$$
$$= 31.584$$

(b) 5.4×0.6

$$
\begin{array}{r}
5.4 \\
\times\ 0.6 \\
\hline
324 \\
000 \\
\hline
03.24
\end{array}
$$

5.4 (1 *decimal place*)
$\times\ 0.6$ (1 *decimal place*)

03.24 (2 *decimal places*)

$5.4 \times 0.6 = 3.24$

OR this can be solved by removing the decimal points first and can be done by using the ideas learned in table *3.1* in chapter *3*.

$$5.4 = 54 \times 10^{-1}$$
$$0.6 = 6 \times 10^{-1}$$
$$5.4 \times 0.6 = 54 \times 10^{-1} \times 6 \times 10^{-1}$$

$$= 54 \times 6 \times 10^{-1} \times 10^{-1}$$
$$= 324 \times 10^{-2}$$
$$= 3.24$$

(c) $0.54 \div 0.06$

move the decimal point two places to the right of a divisor and also a dividend

$$0.54 \div 0.06 = 54 \div 6$$

$$
\begin{array}{r}
9 \\
6\overline{)54} \\
-54 \\
\hline
-\ -
\end{array}
$$

OR this can be solved by removing the decimal points first and can be done by using the ideas learned in table *3.1* in chapter *3*.

$$0.54 \div 0.06 = \frac{0.54}{0.06}$$

$$\frac{0.54}{0.06} = \frac{54 \times 10^{-2}}{6 \times 10^{-2}}$$

$$= \frac{54}{6} \times \frac{10^{-2}}{10^{-2}}$$

$$= 9 \times 10^{(-2 - {}^-2)}$$
$$= 9 \times 10^{(-2 + 2)}$$
$$= 9 \times 10^{0}$$
$$= 9 \times 1$$
$$= 9$$

(d) $12.126 \div 25.8$

move the decimal point one
place to the right of a
divisor and also a dividend

$12.126 \div 25.8 = 121.26 \div 258$

$$
\begin{array}{r}
0.47 \\
258\overline{\smash{)}121.26} \\
-1032 \\
\hline
1806 \\
1806 \\
\hline
\text{- - - -}
\end{array}
$$

OR this can be solved by removing the decimal points first and can be done by using the ideas learned in table *3.1* in chapter *3*.

$$12.126 \div 25.8 = \frac{12.126}{25.8}$$

$$\frac{12.126}{25.8} = \frac{12126 \times 10^{-3}}{258 \times 10^{-1}}$$

$$= \frac{12126}{258} \times \frac{10^{-3}}{10^{-1}}$$

$$= 47 \times 10^{(-3 - ^{-}1)}$$
$$= 47 \times 10^{(-3 + 1)}$$
$$= 47 \times 10^{-2}$$
$$= 0.47$$

(e) $99 \div 0.11$

move the decimal point two
places to the right of a
divisor and also a dividend

$$99.00 \div 0.11 = 9900 \div 11$$

$$
\begin{array}{r}
900 \\
11\overline{)9900} \\
-99 \\
\hline
--\,00
\end{array}
$$

$$99 \div 0.11 = 900$$

OR this can be solved by removing the decimal points first and can be done by using the ideas learned in table *3.1* in chapter *3*.

$$99 \div 0.11 = \frac{99}{0.11}$$

$$\frac{99}{0.11} = \frac{99 \times 10^0}{11 \times 10^{-2}}$$

$$= \frac{99}{11} \times \frac{10^0}{10^{-2}}$$

$$= 9 \times 10^{(0 - {}^-2)}$$
$$= 9 \times 10^2$$
$$= 9 \times 100$$
$$= 900$$

Exercise 4B

(1) Convert the following fractions into decimals and determine whether it is a Terminating Decimal, Repeating Decimal or Irrational number (round off your answers to the nearest a thousandth or three decimal places).

(a) $\dfrac{3}{8}$

(b) $\dfrac{8}{9}$

(c) $\dfrac{9}{13}$

(2) Given two decimal numbers *0.316* and *0.0316*
(i) What is the bigger number?
(ii) Round off the two decimal numbers to hundredth or 2 decimal places.
(3) What is the bigger decimal number ⁻*0.0097* and ⁻*0.097*
(4) Add or Subtract the following Decimals:
(a) 0.989 + 0.011
(b) 4.278 + 32.2375 + 3.6845
(c) 25.62 – 9.42
(d) 12 – 0.02
(5) Multiply or Divide the following Decimals:
(a) 4.2 × 7.52
(b) 5.4 × 0.6
(c) 0.54 ÷ 0.06
(d) 12.126 ÷ 25.8
(e) 99 ÷ 0.11
(6) Evaluate:
(i) 0.18 + 0.081 ÷ 0.9 × 2.0

(ii) $60.0 - 2.52 \div 0.042$

(iii) $0.2 + 0.2 - 0.2 \times 0.2 \div 0.2$

CHAPTER 5

PERCENTAGES

A percent is a fraction with *100* as a denominator
(% is a symbol for per hundred)

$$5\% = \frac{5}{100}$$

$$100\% = \frac{100}{100} = 1$$

In chapter 4 we have seen how to convert a fraction to a decimal by dividing, now we can also convert a decimal into a fraction, for example:

Convert *0.25* into a fraction

Solution

$$0.25_\curvearrowright$$

Have two decimal places and the point moving to the right

$$25.0 \times 10^{-2} = 25.0 \times \frac{1}{10^2}$$

$$= 25 \times \frac{1}{100}$$

$$= \frac{\cancel{25}^1}{\cancel{100}^4}$$

$$= \frac{1}{4}$$

5-1 Converting a number to a percent
Example 5.1

(a) 8

$$8 \times 100\% = 800\%$$

(b) 0.05

Point moves 2 places to the right when multiplying by *100*.

$$0.05 \times 100\% = 5\%$$

OR $\quad 0.05 \times 100\% = 5.0 \times 10^{-2} \times 10^{2}\%$
$$= 5.0 \times 10^{(-2 + 2)}\%$$
$$= 5.0 \times 10^{0}\%$$
$$= 5.0 \times 1\%$$
$$= 5\%$$

(c) $\dfrac{1}{5}$

$$= \frac{1}{5} \times 100\%$$

$$= \frac{1}{5} \times \cancel{100}^{20}\%$$

$$= 20\%$$

5-2 Converting a percent to a fraction
Example 5.2

(a) 80%

$$\frac{80}{100} = \frac{\cancel{80}^{4}}{\cancel{100}^{5}}$$

$$= \frac{4}{5}$$

(b) 110%

$$\frac{110}{100} = \frac{11}{10}$$

(c) 0.5%

$$\frac{0.5}{100} = 0.5 \div 100$$

$$= \frac{5}{10} \div 100$$

$$= \frac{5}{10} \times \frac{1}{100}$$

$$= \frac{5^1}{\cancel{1000}^{200}}$$

$$= \frac{1}{200}$$

5-3 Converting a percent to a decimal
Example 5.3
(a) 20%

$$\frac{20}{100} = \frac{2}{10}$$
$$= 0.2$$

OR

$$\frac{20}{100} = 20 \times \frac{1}{100}$$
$$= 20.0 \times 10^{-2}$$
$$= 0.2$$

(b) 0.9%

$$\frac{0.9}{100} = 0.009$$

OR $$\frac{0.9}{100} = 0.9 \times \frac{1}{100}$$
$$= 0.9 \times 10^{-2}$$
$$= 00.9 \times 10^{-2}$$
$$= 0.009$$

5-4 Computing a percent of the given number
Example 5.4
(a) Compute 50% of 900

Solution

$$\frac{50}{100^1} \times 900^9 = 450$$

(b) What percent of *250* is *40?*

Solution

Let the percent be x
$$x\% \text{ of } 250 = 40$$
$$\frac{x}{100^2} \times 250^5 = 40$$
$$\frac{5x}{2} = 40$$

Multiply both sides by 2
$$2 \times \frac{5x}{2} = 40 \times 2$$
$$5x = 80$$

Divide both sides by 5
$$\frac{5x}{5} = \frac{80^{16}}{5}$$
$$x = 16$$

Therefore ($\therefore$) the percent is *16%*

(c) Twenty percent of five hundred students in Williams high school are boys. Find the number of boys in that school.

Solution

Boys → 20% of 500

$$\frac{20}{\underset{1}{\cancel{100}}} \times \cancel{500}^{5}$$

$$20 \times 5 = 100$$

∴ There are *100* boys in that school.

Exercise 5A

1(a) Convert *0.125* into a fraction

(b) Convert *0.55* into a fraction

(2) Convert the following numbers into percent

(a) $1\frac{2}{4}$

(b) 20

(c) 0.015

(3) Convert a percent to a fraction

(a) 5%

(b) 75%

(c) $33\frac{1}{3}\%$ (d) 170%

(e) 0.25%

(4) Convert a percent to a decimal

(a) 120%

(b) 0.175%

5(a) Compute *75%* of *888*

(b) Compute *120%* of *1000*

6(a) What percent of *70* is *32.2?*

(b) What percent of *125* is $\frac{1}{4}$?

(7) In a certain class *60%* are girls. Given that the number of students are *210*, how many boys are in this class?

(8) Thirty percent of seven hundred students in a certain school are boys. Find the number of boys in that school.

(9) In a basket there are *5* green balls and *10* red balls. What percent of all the balls in the basket are red? (Give your answer to the nearest hundredth).

SOLUTIONS FOR EXERCISE 5A

1(a) $0.1\overset{\frown}{25}$

has three decimal places and the point moving to the right

$$125.0 \times 10^{-3} = 125.0 \times \frac{1}{10^3}$$

$$= 125 \times \frac{1}{1000}$$

$$= \frac{\overset{1}{\cancel{125}}}{\underset{8}{\cancel{1000}}}$$

$$= \frac{1}{8}$$

(b) $0.\overset{\frown}{55}$

Have two decimal places and the point moving to the right

$$55.0 \times 10^{-2} = 55.0 \times \frac{1}{10^2}$$

$$= 55 \times \frac{1}{100}$$

$$= \frac{\overset{11}{\cancel{55}}}{\underset{20}{\cancel{100}}}$$

$$= \frac{11}{20}$$

2(a) $$1\frac{2}{4} = \frac{6}{4} \times 100\%$$

$$= \frac{6}{4^1} \times \cancel{100}^{25}\%$$

$$= 150\%$$

(b) $$20 \times 100\% = 2000\%$$

(c) 0.015

$$0.015 \times 100\% = 1.5\%$$

Point moves 2 places to the right when multiplied by *100*.

OR $0.015 \times 100\% = 1.5 \times 10^{-2} \times 10^2\%$
$$= 1.5 \times 10^{(-2 + 2)}\%$$
$$= 1.5 \times 10^0\%$$
$$= 1.5 \times 1\%$$
$$= 1.5\%$$

3(a) $$5\% = \frac{5}{100}$$

$$= \frac{\cancel{5}^1}{\cancel{100}^{20}}$$

$$= \frac{1}{20}$$

(b) $$75\% = \frac{75}{100}$$

$$= \frac{\cancel{75}^3}{\cancel{100}^4}$$

$$= \frac{3}{4}$$

(c)
$$33\frac{1}{3}\% = \frac{100}{3}\%$$

$$= \frac{100}{3 \times 100}$$

$$= \frac{1}{3}$$

(d)
$$170\% = \frac{170}{100}$$

$$= \frac{17}{10}$$

(e)
$$0.25\% = \frac{0.25}{100}$$

$$= 0.25 \div 100$$

$$= \frac{25}{100} \div 100$$

$$= \frac{25}{100} \times \frac{1}{100}$$

$$= \frac{25}{10000}$$

$$= \frac{1}{400}$$

4(a)
$$120\% = \frac{120}{100}$$

$$= 1.2$$

Decimal point moves 2 places to the left when dividing by *100*.

$$\text{OR} \qquad \frac{120}{100} = 120 \times \frac{1}{100}$$

$$= 120 \times 10^{-2}$$

$$= 120.0 \times 10^{-2}$$

$$= 1.2$$

(b) $\qquad 0.175\% = \dfrac{0.175}{100}$

$$= 0.00175$$

Point moves 2 places to the left when dividing by *100*.

$$\text{OR} \qquad \frac{0.175}{100} = 0.175 \times \frac{1}{100}$$

$$= 0.175 \times 10^{-2}$$

$$= 00.175 \times 10^{-2}$$

$$= 0.00175$$

5(a) $\qquad 75\% \text{ of } 888 = \dfrac{\cancel{75}^{3}}{\cancel{100}^{4}} \times \cancel{888}^{222}$

$$= 3 \times 222$$

$$= 666$$

(b) 120% of $1000 = \dfrac{120}{\cancel{100}} \times \cancel{1000}$

$$= 120 \times 10$$
$$= 1200$$

6(a) Per cent of 70 is 32.2

Let the percent be x

$$x\% \text{ of } 70 = 32.2$$

$$\frac{x}{\cancel{100}^{20}} \times \cancel{70}^{14} = 32.2$$

$$\frac{14x}{20} = 32.2$$

Multiply both sides by 20

$$\cancel{20} \times \frac{14x}{\cancel{20}} = 32.2 \times 20$$

$$14x = 644$$

Divide both sides by 14

$$\frac{\cancel{14}x}{\cancel{14}} = \frac{\cancel{644}^{46}}{\cancel{14}}$$

$$x = 46$$

Therefore ($\therefore$) the percent is 46%

(b) Percent of 125 is $\dfrac{1}{4}$

Let the percent be x

$$x\% \text{ of } 125 = \frac{1}{4}$$

$$\frac{x}{\cancel{100}^{4}} \times \cancel{125}^{5} = \frac{1}{4}$$

95

$$\frac{5x}{4} = \frac{1}{4}$$

Multiply both sides by 4

$$4 \times \frac{5x}{4} = \frac{1}{4} \times 4$$

$$5x = 1$$

Divide both sides by 5

$$\frac{5x}{5} = \frac{1}{5}$$

$$x = \frac{1}{5}$$

Therefore the percent is $\frac{1}{5}\%$

(7) Girls → 60% of 210

$$\frac{60}{100} \times 210 = 126 \text{ girls}$$

Boys → 210 − 126 = 84 boys

(8) Boys → 30% of 700

$$\frac{30}{100^1} \times 700^7 = 30 \times 7$$

$$= 210$$

∴ There are *210* boys in that school.

(9) *5* green balls
10 red balls
All the balls (5 + 10) = 15

$$\text{Percent of Red balls} = \frac{10^2}{15^3} \times 100\%$$

$$= 0.6666\ldots \times 100\%$$

$$= 0.\overset{\curvearrowright\!\downarrow}{6}66\ldots \times 10^2\%$$

$$= 66.66\%$$

Exercise 5B

1(a) Convert *0.96* into a fraction

(b) Convert *0.9375* into a fraction

(2) Convert the following numbers into percent

(a) $2\dfrac{3}{5}$

(b) 0.456

(c) 100

(3) Convert a percent to a fraction

(a) 35%

(b) $9\dfrac{1}{11}\%$

(c) $16\dfrac{2}{3}\%$

(d) 550%

(e) 0.75%

(4) Convert a percent to a decimal

(a) 250%

(b) 0.195%

(c) $\dfrac{1}{2}\%$

(d) $33\dfrac{1}{3}\%$

(e) 165%

5(a) Compute *60%* of *250*

(b) Compute *96%* of *0.25*

(c) Compute *30%* of $3\frac{1}{3}$

6(a) What percent of *111* is *5.55?*

(b) What percent of *400* is $\frac{2}{5}$?

(c) What percent of *250* is *5?*

(7) In a certain class *80%* are boys. Given that the number of students are *120*, how many girls are in this class?

(8) Forty five percent of four hundred students in a certain school are boys. Find the number of girls in that school.

(9) In a basket there are *8* green balls and *12* red balls. What percent of all the balls in the basket are red?

CHAPTER 6

RADICALS

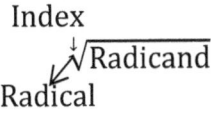

Index
↓
$\sqrt[]{\text{Radicand}}$
Radical

$$\sqrt[3]{8} = 2$$

3 is the index
8 is the radicand
2 is the result since $2 \times 2 \times 2 = 8$

6-1 Calculator approximation of Radicals
Example 6.1
(a) Give a calculator approximation of $\sqrt{246}$ to the nearest tenth.

$$\sqrt{246} = 15.6843871....$$
$$= 15.7$$

(b) Give a calculator approximation of $\sqrt{45}$ to nearest tenth.
$$\sqrt{45} = 6.708203932....$$
$$= 6.7$$

6-2 Laws of radicals
Example 6.2
(a) $\qquad \sqrt[m]{a^m} = a$

if m is an odd positive integer

$$\sqrt[3]{5^3} = 5$$

(b)
$$\sqrt[m]{a^m} = |a|$$
$$= \pm a$$

if m is an even positive integer

$$\sqrt[2]{6^2} = |6|$$
$$= \pm 6$$

Answers are ⁺6 and ⁻6

(c)
$$a^{\frac{1}{m}} = \sqrt[m]{a}$$
for all real numbers

$$9^{\frac{1}{2}} = \sqrt[2]{9}$$
$$= \sqrt{9}$$
$$= 3$$

(d)
$$a^{\frac{n}{m}} = \left(a^{\frac{1}{m}} \right)^n$$

$$4^{\frac{3}{2}} = (4^{\frac{1}{2}})^3$$
$$= (\sqrt{4})^3$$
$$= 2^3$$
$$= 8$$

(e)
$$a^{\frac{-n}{m}} = \frac{1}{a^{\frac{n}{m}}} \qquad a \neq 0$$

$$8^{\frac{-2}{3}} = \frac{1}{8^{\frac{2}{3}}}$$

$$= \frac{1}{(\sqrt[3]{8})^2}$$

$$= \frac{1}{2^2}$$

$$= \frac{1}{4}$$

(f) $\sqrt[m]{a} * \sqrt[m]{b} = \sqrt[m]{ab}$

$$\sqrt{2} * \sqrt{3} = \sqrt{2 \times 3} = \sqrt{6}$$

(g) $\dfrac{\sqrt[m]{a}}{\sqrt[m]{b}} = \sqrt[m]{\dfrac{a}{b}}$

$$\frac{\sqrt{2}}{\sqrt{8}} = \sqrt{\frac{2}{8}}$$

$$= \sqrt{\frac{1}{4}}$$

$$= \frac{1}{2}$$

(h) $i^2 = {}^{-}1, \quad i = \sqrt{{}^{-}1}$

(i)
$$i^4 = i^2 * i^2$$
$$= {}^{-}1 \times {}^{-}1$$
$$= 1$$

(ii)
$$\sqrt{{}^{-}9} = \sqrt{{}^{-}1 * 9}$$
$$= \sqrt{i^2 * 9}$$
$$= i\sqrt{9}$$
$$= 3i$$

(iii)
$$\sqrt{^-7} = \sqrt{^-1 * 7}$$
$$= \sqrt{i^2 * 7}$$
$$= i\sqrt{7}$$

6-3 Rationalize a denominator
Examples 6.3 simplify:

(a)
$$\frac{10}{\sqrt{5}} = \frac{10}{\sqrt{5}} \times \frac{\sqrt{5}}{\sqrt{5}}$$

$$= \frac{10\sqrt{5}}{\sqrt{5 \times 5}}$$

$$= \frac{10\sqrt{5}}{\sqrt{5^2}}$$

$$= \frac{10\sqrt{5}}{5}$$

$$= \frac{\cancel{10}^2\sqrt{5}}{\cancel{5}^1}$$

$$= 2\sqrt{5}$$

OR use exponents $\dfrac{10}{\sqrt{5}} = \dfrac{10}{\sqrt{5}} \times \dfrac{\sqrt{5}}{\sqrt{5}}$

$$= \frac{10\sqrt{5}}{5^{\frac{1}{2}} \times 5^{\frac{1}{2}}}$$

$$= \frac{10\sqrt{5}}{5^{\frac{1}{2}+\frac{1}{2}}}$$

$$= \frac{10\sqrt{5}}{5^1}$$

$$= \frac{\cancel{10}^2\sqrt{5}}{\cancel{5}^1}$$

$$= 2\sqrt{5}$$

(b)
$$\frac{1+i}{1-i} = \frac{(1+i)(1+i)}{(1-i)(1+i)}$$

$$= \frac{1+i+i+i^2}{1+i-i-i^2}$$

$$= \frac{1+2i+i^2}{1-i^2}$$

But $i = \sqrt{-1}, \quad i^2 = {}^-1$
$$= \frac{1+2i-1}{1-{}^-1}$$

$$= \frac{2i}{1+1}$$

$$= \frac{2i}{2}$$

$$= i = \sqrt{-1}$$

6-4 Adding or subtracting radicals
Example 6.4
(a) $\qquad 4\sqrt{2} + 6\sqrt{2} = \sqrt{2}(4 + 6)$
$\qquad \sqrt{2}$ is common, factorized out of the brackets
$$= 10\sqrt{2}$$

$$\text{OR} \qquad \text{Let } p = \sqrt{2}$$
$$4p + 6p = 10p$$
$$\text{But } p = \sqrt{2}$$
$$= 10\sqrt{2}$$

(b) $3\sqrt{3} + 4\sqrt{2} - \sqrt{18} = 3\sqrt{3} + 4\sqrt{2} - \sqrt{9 \times 2}$
$$= 3\sqrt{3} + 4\sqrt{2} - \sqrt{3^2 \times 2}$$
$$= 3\sqrt{3} + 4\sqrt{2} - \sqrt{3^2}\sqrt{2}$$

From $\sqrt[m]{a^m} = a$, $\sqrt{3^2} = 3$

$$= 3\sqrt{3} + 4\sqrt{2} - 3\sqrt{2}$$
$$= 3\sqrt{3} + \sqrt{2}(4 - 3)$$
$$= 3\sqrt{3} + \sqrt{2}$$

6-5 Multiplying or dividing radicals
Example 6.5

(a) $\qquad 4\sqrt{8} * \sqrt{2} = 4\sqrt{4 \times 2} * \sqrt{2}$
$$= 4\sqrt{2^2 \times 2} * \sqrt{2}$$
$$= 4 * 2\sqrt{2} * \sqrt{2}$$
$$= 4 * 2\sqrt{2 \times 2}$$
$$= 8\sqrt{2^2}$$
$$= 8 \times 2$$
$$= 16$$

(b) $\qquad 4\sqrt{12} \div \sqrt{3} = 4\sqrt{4 * 3} \div \sqrt{3}$
$$= 4\sqrt{2^2 * 3} \div \sqrt{3}$$
$$= 4 * 2\sqrt{3} \div \sqrt{3}$$
$$= \frac{8\sqrt{3}}{\sqrt{3}}$$
$$= 8$$

(c) $\qquad \sqrt[3]{27} \times \sqrt[3]{8a^3} = \sqrt[3]{3^3} \times \sqrt[3]{2^3 a^3}$
$$= \sqrt[3]{3^3} \times \sqrt[3]{(2a)^3}$$

$$= 3 \times (2a)$$
$$= 6a$$

6-6 Simplify radicals
Examples 6.6

(a)

$$(7\sqrt{5} + 4)(2\sqrt{5} - 1)$$
$$= 7\sqrt{5} * 2\sqrt{5} - 7\sqrt{5} * 1 + 4 * 2\sqrt{5} - 4 * 1$$
$$= 14\sqrt{5}\sqrt{5} - 7\sqrt{5} + 8\sqrt{5} - 4$$
$$= 14 * 5 - 7\sqrt{5} + 8\sqrt{5} - 4$$
$$= 70 + \sqrt{5}(^-7 + 8) - 4$$
$$= 66 + \sqrt{5}$$

(b)

$$\frac{16}{\sqrt{5} + \sqrt{3}} = \frac{16}{\sqrt{5} + \sqrt{3}} \times \frac{\sqrt{5} - \sqrt{3}}{\sqrt{5} - \sqrt{3}}$$

$$= \frac{16(\sqrt{5} - \sqrt{3})}{\sqrt{5}\,\sqrt{5} - \sqrt{5}\sqrt{3} + \sqrt{5}\sqrt{3} - \sqrt{3}\sqrt{3}}$$

$$= \frac{16(\sqrt{5} - \sqrt{3})}{5 - 3}$$

$$= \frac{\cancel{16}^8(\sqrt{5} - \sqrt{3})}{\cancel{2}}$$

$$= 8(\sqrt{5} - \sqrt{3})$$

Exercise 6A

(1) Give a calculator approximation of $\sqrt{130.250}$ to the nearest thousandth.

Simplify each expression:

(2) $3\sqrt{20} - 5\sqrt{80} + 4\sqrt{320}$

(3) $3\sqrt{12} + \sqrt{20} + \sqrt{27} + \sqrt{500}$

(4) $3\sqrt{2} + 7\sqrt{2} - 2\sqrt{2}$

(5) $\sqrt{45a} - \sqrt{5a}$

(6) $\sqrt[3]{16a^3}$

(7) $64^{\frac{1}{2}}$

(8) $25^{\frac{3}{2}}$

(9) $64^{\frac{-2}{3}}$

(10) $(a^5 b^3 c^6)^{\frac{1}{3}}$

(11) $\sqrt{\dfrac{8a}{b}} * \sqrt{\dfrac{b}{4}}$

(12) $\sqrt{\dfrac{8x^2 y^3}{z^2}}$

(13) $8x^{12}y^4 \div 2x^9 y^2$

(14) $\sqrt[4]{16a^8 b^{24} c^2}$

(15) $2\sqrt{20} * 3\sqrt{45}$

(16) $\sqrt[3]{9} * \sqrt{9}$

(17) $2\sqrt{250} \div 5\sqrt{40}$

(18) $\dfrac{5^{\frac{2}{5}}x^{\frac{2}{5}}y^{\frac{-1}{4}}}{5^{\frac{1}{10}}x^{\frac{-8}{5}}y^{\frac{7}{4}}}$

(19) $(5\sqrt{7} + 2)(2\sqrt{7} - 1)$

(20) $\dfrac{-20\sqrt[3]{5}}{\sqrt[3]{40}}$

(21) $\dfrac{2}{\sqrt{5} + \sqrt{7}}$

(22) i^{10}

(23) $\dfrac{2 + i}{3 - i}$

SOLUTIONS FOR EXERCISE 6A

(1) $\sqrt{130.250} = 11.41271221\ldots\ldots$

$\qquad \approx 11.413$

(2) $\qquad 3\sqrt{20} - 5\sqrt{80} + 4\sqrt{320}$

$= 3\sqrt{4 \times 5} - 5\sqrt{16 \times 5} + 4\sqrt{64 \times 5}$

$= 3\sqrt{2^2 \times 5} - 5\sqrt{4^2 \times 5} + 4\sqrt{8^2 \times 5}$

$= 3 * 2\sqrt{5} - 5 * 4\sqrt{5} + 4 * 8\sqrt{5}$

$= 6\sqrt{5} - 20\sqrt{5} + 32\sqrt{5}$

$= \sqrt{5}(6 - 20 + 32)$

$= 18\sqrt{5}$

(3) $\qquad 3\sqrt{12} + \sqrt{20} + \sqrt{27} + \sqrt{500}$

$= 3\sqrt{4 \times 3} + \sqrt{4 \times 5} + \sqrt{9 \times 3} + \sqrt{100 \times 5}$

$= 3\sqrt{2^2 \times 3} + \sqrt{2^2 \times 5} + \sqrt{3^2 \times 3} + \sqrt{10^2 \times 5}$

$= 3 * 2\sqrt{3} + 2\sqrt{5} + 3\sqrt{3} + 10\sqrt{5}$

$= 6\sqrt{3} + 2\sqrt{5} + 3\sqrt{3} + 10\sqrt{5}$

$= 6\sqrt{3} + 3\sqrt{3} + 2\sqrt{5} + 10\sqrt{5}$

$= 9\sqrt{3} + 12\sqrt{5}$

$= 3(3\sqrt{3} + 4\sqrt{5})$

(4) $3\sqrt{2} + 7\sqrt{2} - 2\sqrt{2}$

$= \sqrt{2}(3 + 7 - 2)$

$= 8\sqrt{2}$

(5) $\sqrt{45a} - \sqrt{5a}$

$= \sqrt{9 \times 5a} - \sqrt{5a}$

$= \sqrt{3^2 \times 5a} - \sqrt{5a}$

$= 3\sqrt{5a} - \sqrt{5a}$

$= 2\sqrt{5a}$

(6) $\sqrt[3]{16a^3}$

$= \sqrt[3]{8 * 2a^3}$

$= \sqrt[3]{2^3 * 2a^3}$

$= 2a\sqrt[3]{2}$

(7) $64^{\frac{1}{2}}$

$= \sqrt[2]{64}$

$= \sqrt{64}$

$= \sqrt{8^2}$

$= 8$

(8) $25^{\frac{3}{2}}$

$= (25^{\frac{1}{2}})^3$

$$= (\sqrt{25})^3$$
$$= (\sqrt{5^2})^3$$
$$= 5^3$$
$$= 5 \times 5 \times 5$$
$$= 125$$

(9) $\qquad 64^{\frac{-2}{3}}$

$$= \frac{1}{64^{\frac{2}{3}}}$$

$$= \frac{1}{(\sqrt[3]{64})^2}$$

$$= \frac{1}{(\sqrt[3]{4^3})^2}$$

$$= \frac{1}{4^2}$$

$$= \frac{1}{16}$$

(10) $\qquad (a^5 b^3 c^6)^{\frac{1}{3}}$

$$= a^{(5 \times \frac{1}{3})} b^{(3 \times \frac{1}{3})} c^{(6 \times \frac{1}{3})}$$

$$= a^{\frac{5}{3}} b^1 c^2$$

$$= \sqrt[3]{a^5}\, bc^2$$

$$= \sqrt[3]{a^3 * a^2}\, bc^2$$

$$= abc^2 \sqrt[3]{a^2}$$

(11) $$\sqrt{\frac{8a}{b}} * \sqrt{\frac{b}{4}}$$

$$= \sqrt{\frac{8ab}{4b}}$$

$$= \sqrt{\frac{4 \times 2a\cancel{b}}{4\cancel{b}}}$$

$$= \sqrt{2a}$$

(12) $$\sqrt{\frac{8x^2y^3}{z^2}}$$

$$= \sqrt{\frac{2^3x^2y^3}{z^2}}$$

$$= \sqrt{\frac{2*2^2x^2y^2*y}{z^2}}$$

$$= \sqrt{\frac{(2xy)^2*2y}{z^2}}$$

$$= \frac{2xy}{z}\sqrt{2y}$$

(13) $$8x^{12}y^4 \div 2x^9y^2$$

$$= 2^3x^{12}y^4 \div 2x^9y^2$$

$$= 2^{(3-1)}x^{(12-9)}y^{4-2)}$$

$$= 2^2x^3y^2$$

$$= 2^2y^2x^3$$

$$= (2y)^2 x^3$$

(14) $\sqrt[4]{16a^8b^{24}c^2}$

$$= (16a^8b^{24}c^2)^{\frac{1}{4}}$$

$$= (4^2 a^8 b^{24} c^2)^{\frac{1}{4}}$$

$$= 4^{(2 \times \frac{1}{4})} a^{(8 \times \frac{1}{4})} b^{(24 \times \frac{1}{4})} c^{(2 \times \frac{1}{4})}$$

$$= 4^{\frac{1}{2}} a^2 b^6 c^{\frac{1}{2}}$$

$$= \sqrt{4}\, a^2 b^6 \sqrt{c}$$

$$= \sqrt{4}\, \sqrt{c}\, a^2 b^6$$

$$= \sqrt{2^2}\, \sqrt{c}\, a^2 b^6$$

$$= 2\sqrt{c}\, (ab^3)^2$$

(15) $2\sqrt{20} * 3\sqrt{45}$

$$= 2\sqrt{4 \times 5} * 3\sqrt{9 \times 5}$$

$$= 2\sqrt{2^2 \times 5} * 3\sqrt{3^2 \times 5}$$

$$= 2 * 2\sqrt{5} * 3 * 3\sqrt{5}$$

$$= 4\sqrt{5} * 9\sqrt{5}$$

$$= (4 \times 9)(\sqrt{5} \times \sqrt{5})$$

$$= 36 \times 5$$

$$= 180$$

(16) $\sqrt[3]{9} * \sqrt{9}$

$$= \sqrt[3]{9} * \sqrt{3^2}$$

$$= \sqrt[3]{9} * 3$$

$$= 3\sqrt[3]{9}$$

(17)
$$2\sqrt{250} \div 5\sqrt{40}$$

$$= 2\sqrt{25 \times 10} \div 5\sqrt{4 \times 10}$$

$$= \frac{2\sqrt{25 \times 10}}{5\sqrt{4 \times 10}}$$

$$= \frac{2\sqrt{5^2 \times 10}}{5\sqrt{2^2 \times 10}}$$

$$= \frac{2*5\sqrt{10}}{5*2\sqrt{10}}$$

$$= \frac{10\sqrt{10}}{10\sqrt{10}}$$

$$= 1$$

(18)
$$\frac{5^{\frac{2}{5}} x^{\frac{2}{5}} y^{\frac{-1}{4}}}{5^{\frac{1}{10}} x^{\frac{-8}{5}} y^{\frac{7}{4}}}$$

$$= 5^{\left(\frac{2}{5} - \frac{1}{10}\right)} x^{\left(\frac{2}{5} - \frac{-8}{5}\right)} y^{\left(\frac{-1}{4} - \frac{7}{4}\right)}$$

$$= 5^{\frac{4-1}{10}} x^{\frac{2+8}{5}} y^{\frac{-1-7}{4}}$$

$$= 5^{\frac{3}{10}} x^{\frac{10}{5}} y^{\frac{-8}{4}}$$

$$= 5^{\frac{3}{10}} x^2 y^{-2}$$

$$= \left(\sqrt[10]{5}\right)^3 x^2 y^{-2} \qquad \text{but } y^{-2} = \frac{1}{y^2}$$

$$= \left(\sqrt[10]{5} \right)^3 \frac{x^2}{y^2}$$

$$= \left(\sqrt[10]{5} \right)^3 \left(\frac{x}{y} \right)^2$$

(19) $(5\sqrt{7} + 2)(2\sqrt{7} - 1)$

$= 5\sqrt{7} * 2\sqrt{7} + 5\sqrt{7} * \text{-}1 + 2 * 2\sqrt{7} + 2 * \text{-}1$

$= 10 * 7 - 5\sqrt{7} + 4\sqrt{7} - 2$

$= 70 - \sqrt{7} - 2$

$= 68 - \sqrt{7}$

(20) $\dfrac{-20\sqrt[3]{5}}{\sqrt[3]{40}}$

$= \dfrac{-20\sqrt[3]{5}}{\sqrt[3]{8 \times 5}}$

$= \dfrac{-20\sqrt[3]{5}}{\sqrt[3]{2^3 \times 5}}$

$= \dfrac{-20\sqrt[3]{5}}{\sqrt[3]{2^3}\,\sqrt[3]{5}}$

$= \dfrac{-20}{2}$

$= \text{-}10$

(21) $\dfrac{2}{\sqrt{5} + \sqrt{7}}$

$= \dfrac{2}{\sqrt{5} + \sqrt{7}} \times \dfrac{\sqrt{5} - \sqrt{7}}{\sqrt{5} - \sqrt{7}}$

$$= \frac{2(\sqrt{5} - \sqrt{7})}{(\sqrt{5} + \sqrt{7})(\sqrt{5} - \sqrt{7})}$$

$$= \frac{2(\sqrt{5} - \sqrt{7})}{\sqrt{5} * \sqrt{5} - \sqrt{5}\sqrt{7} + \sqrt{5}\sqrt{7} - \sqrt{7}\sqrt{7}}$$

$$= \frac{2(\sqrt{5} - \sqrt{7})}{5 - 7}$$

$$= \frac{2(\sqrt{5} - \sqrt{7})}{-2}$$

$$= \frac{\cancel{2}(\sqrt{5} - \sqrt{7})}{-\cancel{2}}$$
$$= - (\sqrt{5} - \sqrt{7})$$

(22) i^{10}

$= (i^2)^5$

$= (-1)^5$

$= -1$

(23) $\dfrac{2 + i}{3 - i}$

$$= \frac{2 + i}{3 - i} \times \frac{3 + i}{3 + i}$$

$$= \frac{(2+i)(3+i)}{(3-i)(3+i)}$$

$$= \frac{6+2i+3i+i^2}{9+3i-3i-i^2}$$

$$= \frac{6+5i+i^2}{9-i^2}$$

$$= \frac{6+5i+{}^-1}{9-{}^-1}$$

$$= \frac{6+5i-1}{9+1}$$

$$= \frac{5+5i}{10}$$

$$= \frac{\cancel{5}(1+i)}{\cancel{10}^2}$$

$$= \frac{1+i}{2}$$

$$= \frac{1+\sqrt{-1}}{2}$$

Exercise 6B

(1) Give a calculator approximation of $\sqrt{150.50}$ to the nearest tenth.

Simplify each expression:

(2) $4\sqrt{3} + 9\sqrt{3} - 2\sqrt{3}$

(3) $2\sqrt{20} - 3\sqrt{12} + \sqrt{125} + \sqrt{300}$

(4) $\sqrt{32} - \sqrt{27} + \sqrt{75}$

(5) $\sqrt{63a} - 4\sqrt{7a}$

(6) $\sqrt[3]{24b^3}$

(7) $\sqrt[3]{54y^4}$

(8) $49^{\frac{1}{2}}$

(9) $4^{\frac{5}{2}}$

(10) $125^{\frac{-2}{3}}$

(11) $(x^4 y^3 z^6)^{\frac{1}{3}}$

(12) $\sqrt{\dfrac{27x}{y}} * \sqrt{\dfrac{xy}{9}}$

(13) $\sqrt{\dfrac{18a^2 b^3 c}{c^3}}$

(14) $12x^{11}y^7 \div 3x^7 y^3$

(15) $\sqrt[4]{81a^{12}b^2 c^8}$

(16) $3\sqrt{48} * \sqrt{300}$

(17) $\sqrt[3]{8} * \sqrt{8}$

(18) $3\sqrt{490} \div 14\sqrt{40}$

(19) $\dfrac{6^{\frac{2}{3}}x^{\frac{1}{3}}y^{\frac{-1}{4}}}{6^{\frac{-1}{3}}x^{\frac{-2}{3}}y^{\frac{3}{4}}}$

(20) $(\sqrt{6} + 3)(3\sqrt{6} - 1)$

(21) $\dfrac{-\sqrt[3]{81}}{6}$

(22) $\dfrac{2\sqrt{5}}{\sqrt{5} + \sqrt{3}}$

(23) i^{12}

(24) $\dfrac{3 + i}{2 - i}$

CHAPTER 7

LINEAR EQUATION

Linear equation is an equation with one or two variables (x,y) of degree one given in this form $ax + b = c$. where a, b and c are real numbers and $a \neq 0$. For the one with two variables (see chapter *11*) is given in this form $ax + by = c$. where a and b cannot be both equal to zero.

7-1 Solving for x in a linear equation AX + B = C   A $\neq$ 0
Examples 7.1
(a) $2x + 8 = 20$

Solution

$$2x + 8 = 20$$

Subtract *8* from both sides of the equation

$$2x + 8 - 8 = 20 - 8$$
$$2x = 12$$

Divide by *2* on both sides of the equation

$$\frac{2x}{2} = \frac{12}{2}$$

$$\frac{\cancel{2}x}{\cancel{2}} = \frac{\cancel{12}^{6}}{\cancel{2}}$$

$$x = 6$$

(b) $x - 7 = 20 - 2x$

Solution

$$x - 7 = 20 - 2x$$

Add 7 to both sides of the equation

$$x - 7 + 7 = 20 - 2x + 7$$
$$x = 27 - 2x$$

Add $2x$ to both sides

$$x + 2x = 27 - 2x + 2x$$
$$3x = 27$$

Divide both sides by 3

$$\frac{3x}{3} = \frac{27}{3}$$
$$x = 9$$

OR $\quad x - 7 = 20 - 2x$

$\left(\begin{array}{l}\textit{collect like terms together and if one term crosses} \\ \textit{an equal sign, its sign changes to the opposite sign.}\end{array}\right)$

$$x + 2x = 20 + 7$$
$$3x = 27$$

Divide both sides by 3

$$\frac{3x}{3} = \frac{27}{3}$$

$$\frac{\cancel{3}x}{\cancel{3}} = \frac{\cancel{27}^9}{\cancel{3}}$$
$$x = 9$$

(c) $\qquad \dfrac{3}{4}x = 6$

Solution

$$\frac{3}{4}x = 6$$

Multiply both sides by 4

$$4 \times \frac{3}{4}x = 6 \times 4$$
$$3x = 24$$

Divide both sides by 3

$$\frac{\cancel{3}x}{\cancel{3}} = \frac{\cancel{24}^8}{\cancel{3}}$$

$$x = 8$$

OR $\frac{3}{4}x = 6$

Multiply by a reciprocal of a coefficient of x on both sides

$$\frac{4}{\cancel{3}} \times \frac{\cancel{3}}{\cancel{4}}x = \cancel{6}^2 \times \frac{4}{\cancel{3}}$$

$$x = 8$$

(d) $\quad \dfrac{3x + 1}{2} = \dfrac{4x + 8}{4}$

Solution

$$\frac{3x + 1}{2} = \frac{4x + 8}{4}$$

Use cross multiplication

$$\frac{3x + 1}{2} \diagdown \diagup \frac{4x + 8}{4}$$

$2(4x + 8) = 4(3x + 1)$

$8x + 16 = 12x + 4 \hspace{2cm} (i)$

Subtract 16 from both sides

$8x + 16 - 16 = 12x + 4 - 16$

$8x \quad = \quad 12x - 12$

Subtract 12x on both sides

$8x - 12x = 12x - 12x - 12$

$-4x = -12$

Divide by −4 on both sides

$$\frac{-4x}{-4} = \frac{-\cancel{12}^{3}}{-4}$$

$$x = 3$$

$OR\ from\ (i)\ \ 8x + 16 = 12x + 4$
correct like terms together
$$8x - 12x = 4 - 16$$
$$-4x = -12$$
Divide both sides by ⁻4
$$\frac{-4x}{-4} = \frac{-\cancel{12}^{3}}{-4}$$
$$x = 3$$

(e) $\sqrt{x + 6} = 4$

Solution

$$\sqrt{x + 6} = 4$$
Square both sides
$$(\sqrt{x + 6})^2 = (4)^2$$
$$x + 6 = 16$$
Subtract 6 from both sides
$$x + 6 - 6 = 16 - 6$$
$$x = 10$$

(f) $\frac{3}{4}x - 3 = \frac{2}{3}x$

Solution

$$\frac{3}{4}x - 3 = \frac{2}{3}x$$

Add 3 to both sides

$$\frac{3}{4}x - 3 + 3 = \frac{2}{3}x + 3$$

$$\frac{3}{4}x = \frac{2}{3}x + 3$$

Subtract $\frac{2}{3}x$ on both sides

$$\frac{3}{4}x - \frac{2}{3}x = \frac{2}{3}x - \frac{2}{3}x + 3$$

$$\frac{3}{4}x - \frac{2}{3}x = 3$$

$$LCD = 12$$

$$\frac{9x - 8x}{12} = 3$$

Multiply both sides by 12

$$\cancel{12} \times \frac{x}{\cancel{12}} = 3 \times 12$$
$$x = 36$$

(g) $0.5x - 25 = 0.25x$

Solution

$$0.5x - 25 = 0.25x$$
Add 25 to both sides
$$0.5x - 25 + 25 = 0.25x + 25$$
$$0.5x = 0.25x + 25$$
Subtract $0.25x$ on both sides
$$0.5x - 0.25x = 0.25x - 0.25x + 25$$
$$0.25x = 25$$

$$\text{But } 0.25 = \frac{25}{100}$$

$$\frac{25}{100}x = 25$$

Multiply both sides by a reciprocal of a coefficient of x

$$\frac{\cancel{100}}{\cancel{25}} \times \frac{\cancel{25}}{\cancel{100}}x = \cancel{25} \times \frac{100}{\cancel{25}}$$

$$x = 100$$

(h) $\quad \dfrac{3}{4} - \dfrac{x}{7} = \dfrac{13}{28}$

Solution

$$\frac{3}{4} - \frac{x}{7} = \frac{13}{28}$$

LCD of 4 and 7 is 28

$$\frac{21 - 4x}{28} = \frac{13}{28}$$

Multiply both sides by 28

$$\cancel{28} \times \frac{21 - 4x}{\cancel{28}} = \frac{13}{\cancel{28}} \times \cancel{28}$$

$$21 - 4x = 13$$

Subtract 21 on both sides

$$21 - 21 - 4x = 13 - 21$$

$$-4x = -8$$

Divide both sides by −4

$$\frac{-4x}{-4} = \frac{\cancel{-8}^{2}}{-4}$$

$$x = 2$$

(i) $\quad$ Solve $V = \dfrac{1}{3}bh$ for h

Solution

Make h the subject

$$V = \frac{1}{3}bh$$
$$3V = bh$$
$$bh = 3V$$
$$\frac{bh}{b} = \frac{3V}{b}$$
$$h = \frac{3V}{b}$$

(j) Twice the difference between a number and 2 is 18.
Find the number

Solution

Let the number be x

$$2(x - 2) = 18$$
$$2x - 4 = 18$$
$$2x = 18 + 4$$
$$2x = 22$$
$$\frac{2x}{2} = \frac{22}{2}$$
$$x = 11$$
$$\therefore \quad \text{the number is } 11$$

7-2 Solving for x in nonlinear equation $AX^n + B = C$, $A \neq 0$
Examples 7.2
(a) $3x^2 = 300$

Solution

$$3x^2 = 300$$

Divide both sides by 3

$$\frac{\cancel{3}x^2}{\cancel{3}} = \frac{\cancel{300}^{100}}{\cancel{3}}$$

$$x^2 = 100$$

Introduce square roots on both sides

$$\sqrt{x^2} = \sqrt{100}$$
$$\sqrt{x^2} = \sqrt{10^2}$$
$$x = 10$$

(b) $\qquad 1\frac{1}{2}x^3 = 12$

Solution

$$1\frac{1}{2}x^3 = 12$$

$$\frac{3}{2}x^3 = 12$$

Multiply both sides by a reciprocal of a coefficient of x^3

$$\frac{\cancel{2}}{\cancel{3}} \times \frac{3x^3}{\cancel{2}} = \cancel{12}^4 \times \frac{2}{3}$$

$$x^3 = 8$$

Introduce cube roots on both sides

$$\sqrt[3]{x^3} = \sqrt[3]{8}$$
$$\sqrt[3]{x^3} = \sqrt[3]{2^3}$$
$$x = 2$$

7-3 Solving for x in an inequality linear equation

$\text{AX} + \text{B} < \text{C} \quad \text{0r} \quad \text{AX} + \text{B} > \text{C} \quad \text{A} \neq 0$

On a number line $0 < 5$

Multiply both sides by 2

$$2 \times 0 < 5 \times 2$$
$$0 < 10$$

If multiplied both sides by ‾2

$$^-2 \times 0 < 5 \times {}^-2$$
$$0 \not< {}^-10$$
$$\therefore \quad 0 > {}^-10$$

0 is not less than *‾10, 0* is greater than *‾10*. Therefore the inequality sign changes when multiplying or dividing by a negative number.

Examples 7.3
Solve for x in the following inequality

(a) $\qquad\qquad 3x - 6 > 21$

Solution

$$3x - 6 > 21$$
$$3x - 6 + 6 > 21 + 6$$
$$3x \;>\; 27$$
$$\frac{\cancel{3}x}{\cancel{3}} > \frac{\cancel{27}^9}{\cancel{3}}$$
$$x > 9$$

OR a solution set $x = \{10, 11, 12, 13, \ldots\ldots\}$

(b) $\qquad\qquad 1 - \dfrac{2x}{3} \geq \dfrac{1}{3}$

Solution

$$1 - \frac{2x}{3} \geq \frac{1}{3}$$

$$1 - 1 - \frac{2x}{3} \geq \frac{1}{3} - 1$$

$$-\frac{2x}{3} \geq -\frac{2}{3}$$

$$\cancel{3} \times \frac{-2x}{\cancel{3}} \geq \frac{-2}{\cancel{3}} \times \cancel{3}$$
$$^-2x \geq {}^-2$$

$$\frac{^-2x}{^-2} \leq \frac{^-2}{^-2}$$

Inequality sign changes when dividing by a negative number

$$x \leq 1$$

7-4 Inequality and Absolute value

If $|ax + b| > K$ therefore, $ax + b > K$ OR $ax + b < {}^-K$

Example 7.4

Solve $|5x - 4| + 2 \leq 8$

Solution

$$|5x - 4| + 2 \leq 8$$
$$|5x - 4| + 2 - 2 \leq 8 - 2$$
$$|5x - 4| \leq 6$$
$$5x - 4 \leq 6$$
$$5x - 4 + 4 \leq 6 + 4$$
$$5x \leq 10$$
$$\frac{5x}{5} \leq \frac{10}{5}$$

$$x \leq 2$$

OR $$5x - 4 \geq {}^-6$$
$$5x - 4 + 4 \geq {}^-6 + 4$$
$$5x \geq {}^-2$$

$$\frac{5x}{5} \geq \frac{{}^-2}{5}$$

$$x \geq \frac{{}^-2}{5}$$

$$\therefore \frac{{}^-2}{5} \leq x \leq 2$$

Exercise 7A

Solve for x

(1) $4x + 7 = 5$

(2) $10x - 10 = 50 + 7x$

(3) $5(x - 2) = {}^-(2x + 3)$

(4) $\dfrac{4}{5}x = \dfrac{1}{5}$

(5) $\dfrac{2x}{7} - \dfrac{x}{5} = \dfrac{9}{35}$

(6) $\dfrac{19x + 1}{2} = \dfrac{11x + 4}{8}$

(7) $\dfrac{1}{4}x - 3 = {}^-3x$

(8) $0.3x - 84 = 0.09x$

(9) $\sqrt{2x + 3} = 5$

(10) $2x^2 = 288$

(11) $2\dfrac{1}{2}x^3 = 160$

(12) $\dfrac{9}{x + 2} - \dfrac{7}{x - 2} = \dfrac{8}{x^2 - 4}$

(13) $\sqrt{x^2 - 3x + 39} = x + 2$

(14) $x + 2 > 8$ and put your answer on a number line

(15) $\dfrac{{}^-1}{2}x \geq 7$ and put your answer on a number line

(16) $2x - 32 > 3x + 16$

(17) ${}^-2x + 1 \leq x + 16$

(18) $\dfrac{3x}{4} - 3 > 6$

(19) $\sqrt{x + 30} \geq 4$

(20) $|x - 3| = 3$ and put your answer on a number line

(21) $|x + 1| < 5$ and put your answer on a number line

(22) $|10 - 2x| \geq 20$

(23) $|3x - 1| + 4 \leq 18$

(24) Solve $V = \frac{1}{3}\pi r^2 h$ for h

(25) A number increased by 13 is 30. Find the number

(26) 5 less than a number is 25. Find the number

(27) Twice the difference between a number and 5 is 100. Find the number.

(28) Find two consecutive odd numbers whose sum is 24

(29) Find the three consecutive integers whose sum is 126

(30) The sum of two numbers is 33 and their quotient is $\frac{1}{2}$ Find the numbers.

(31) A father had 33 oranges to give to his two sons. The elder son was given twice as much oranges as the younger son. How many oranges did the elder son get?

(32) Two-thirds of all students in a class are girls. There are 20 boys. Find the number of all students in the class.

(33) A third of a number is two less than a quarter of a number. Find the number.

(34) John has one-dollar bills and five-dollar bills in his bag. In his bag there are 44 bills together which has a value of 172. How many one-dollar bills and five-dollar bills are in John's bag?

(35) Which of the following equation has a right solution for the number line?

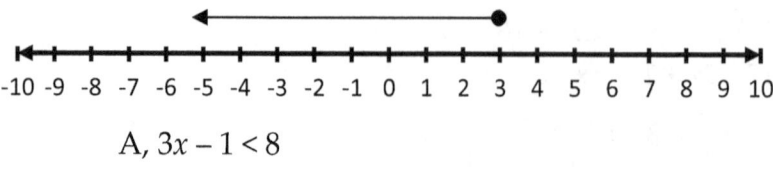

A, $3x - 1 < 8$

B, $3x - 1 > 8$
C, $3x - 1 \leq 8$
D, $3x - 1 \geq 8$

(36) Which of the following equation has a right solution
for the number line?

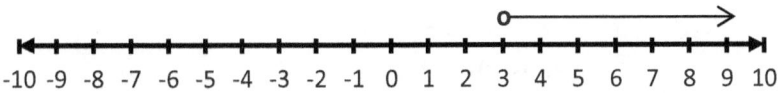

A, $3x - 1 < 8$
B, $3x - 1 > 8$
C, $3x - 1 \leq 8$
D, $3x - 1 \geq 8$

SOLUTIONS FOR EXERCISE 7A

(1)
$$4x + 7 = 5$$
Subtract 7 from both side
$$4x + 7 - 7 = 5 - 7$$
$$4x = {}^-2$$
Divide both sides by 4
$$\frac{4x}{4} = \frac{{}^-2}{4}$$
$$x = \frac{{}^-1}{2}$$

(2)
$$10x - 10 = 50 + 7x$$
Collect like terms together
$$10x - 7x = 50 + 10$$
$$3x = 60$$
Divide both sides by 3
$$\frac{3x}{3} = \frac{60^{20}}{3}$$
$$x = 20$$

(3)
$$5(x - 2) = {}^-(2x + 3)$$
$$5x - 10 = {}^-2x - 3$$
Collect like terms together
$$5x + 2x = 10 - 3$$
$$7x = 7$$
$$x = 1$$

(4)
$$\frac{4}{5}x = \frac{1}{5}$$
Multiply both sides by reciprocal of coefficient of x

$$\frac{5}{4} \times \frac{4}{5}x = \frac{1}{5} \times \frac{5}{4}$$

$$x = \frac{1}{4}$$

(5)

$$\frac{2x}{7} - \frac{x}{5} = \frac{9}{35}$$

$$\frac{5*2x - 7*x}{35} = \frac{9}{35}$$

$$\frac{10x - 7x}{35} = \frac{9}{35}$$

Multiply both sides by 35

$$35 \times \frac{10x - 7x}{35} = \frac{9}{35} \times 35$$

$$10x - 7x = 9$$

$$3x = 9$$

Divide both sides by 3

$$\frac{3x}{3} = \frac{9}{3}$$

$$x = 3$$

(6)

$$\frac{19x + 1}{2} = \frac{11x + 4}{8}$$

Multiply both sides by 8

$$8^4 \times \frac{19x + 1}{2} = \frac{11x + 4}{8} \times 8$$

$$4(19x + 1) = 11x + 4$$

$$76x + 4 = 11x + 4$$

Collect like terms together

133

$$76x - 11x = 4 - 4$$
$$65x = 0$$
Divide both sides by 65
$$x = 0$$

(7)
$$\frac{1}{4}x - 3 = {}^-3x$$

Collect like terms together
$$\frac{1}{4}x + 3x = 3$$

$$\frac{x+12x}{4} = 3$$

Multiply both sides by 4
$$13x = 12$$
Divide both sides by 13
$$x = \frac{12}{13}$$

(8)
$$0.3x - 84 = 0.09x$$

Collect like terms together
$$0.3x - 0.09x = 84$$
$$0.21x = 84$$
$$\frac{21}{100}x = 84$$

Multiply by reciprocal of coefficient of x
$$\frac{100}{21} * \frac{21}{100}x = 84 * \frac{100}{21}$$
$$x = 4 * 100$$
$$x = 400$$

(9)
$$\sqrt{2x + 3} = 5$$

Square both sides

$$\left(\sqrt{2x+3}\right)^2 = 5^2$$
$$2x + 3 = 25$$
$$2x + 3 - 3 = 25 - 3$$
$$2x = 22$$
Divide both sides by 2
$$x = 11$$

(10)
$$2x^2 = 288$$
Divide both sides by 2
$$\frac{2x^2}{2} = \frac{288}{2}$$
$$x^2 = 144$$
Introduce square roots on both sides
$$\sqrt{x^2} = \sqrt{144}$$
$$x = 12$$

(11)
$$2\frac{1}{2}x^3 = 160$$

$$\frac{5}{2}x^3 = 160$$
Multiply both sides by reciprocal of coefficient of x^3
$$\frac{2}{5} \times \frac{5}{2}x^3 = 160 \times \frac{2}{5}$$
$$x^3 = 64$$
$$x^3 = 4^3$$
$$x = 4$$

(12)
$$\frac{9}{x+2} - \frac{7}{x-2} = \frac{8}{x^2-4}$$

$$\frac{9(x-2)-7(x+2)}{(x+2)(x-2)} = \frac{8}{x^2-4}$$

$$\frac{9(x-2)-7(x+2)}{(x+2)(x-2)} = \frac{8}{(x+2)(x-2)}$$

$$9(x-2) - 7(x+2) = 8$$
$$9x - 18 - 7x - 14 = 8$$
$$9x - 7x - 32 = 8$$
$$2x - 32 + 32 = 8 + 32$$
$$2x = 40$$
$$\frac{2x}{2} = \frac{40}{2}$$
$$x = 20$$

(13) $\sqrt{x^2 - 3x + 39} = x + 2$

Square both sides

$$\left(\sqrt{x^2 - 3x + 39}\right)^2 = (x + 2)^2$$
$$x^2 - 3x + 39 = x^2 + 4x + 4$$
$$x^2 - x^2 - 3x - 4x = 4 - 39$$
$$^-7x = {}^-35$$
$$\frac{^-7x}{^-7} = \frac{^-35}{^-7}$$
$$x = 5$$

(14) $$x + 2 > 8$$
$$x + 2 - 2 > 8 - 2$$
$$x > 6$$

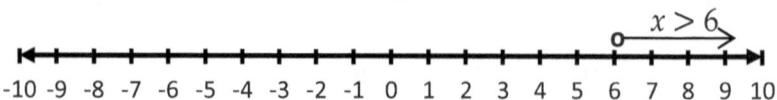

(15) $$\frac{^-1}{2}x \geq 7$$

$$2 \times \frac{^-1}{2}x \geq 7 \times 2$$

$$^-x \geq 14$$

$$x \leq ^-14$$

Sign changed when multiplied by $^-1$

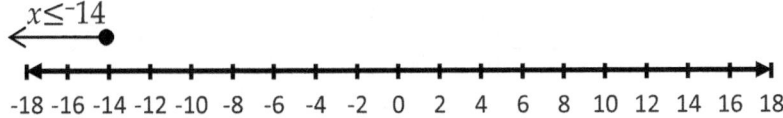

(16) $$2x - 32 > 3x + 16$$

Collect like terms together

$$2x - 3x > 16 + 32$$

$$^-x > 48$$

Multiply or divide both sides by $^-1$

$$x < ^-48$$

(17) $$^-2x + 1 \leq x + 16$$

$$^-2x + 1 - 1 \leq x + 16 - 1$$

$$^-2x - x \leq x - x + 15$$

$$^-3x \leq 15$$

Divide both sides by $^-3$

$$\frac{^-3x}{^-3} \geq \frac{15}{^-3}$$

$$x \geq ^-5$$

(18) $$\frac{3x}{4} - 3 > 6$$

Multiply each term by 4

$$\frac{3x}{4} \times 4 - (3 \times 4) > (6 \times 4)$$

$$3x - 12 > 24$$

$$3x > 24 + 12$$

$$3x > 36$$

$$\frac{3x}{3} > \frac{36}{3}$$
$$x > 12$$

(19)

$$\sqrt{x + 30} \geq 4$$

Square both sides

$$\left(\sqrt{x + 30}\right)^2 \geq 4^2$$
$$x + 30 \geq 16$$
$$x \geq 16 - 30$$
$$x \geq {}^-14$$

(20)

$$|x - 3| = 3$$
$$x - 3 = \pm 3$$
$$x - 3 = 3$$
$$x = 3 + 3$$
$$x = 6$$

OR $\quad x - 3 = {}^-3$
$$x = {}^-3 + 3$$
$$x = 0$$

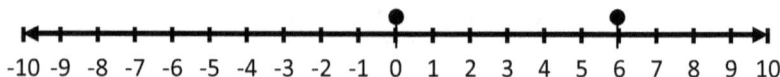

(21)

$$|x + 1| < 5$$
$$x + 1 < 5$$
$$x < 5 - 1$$
$$x < 4$$

OR $\quad x + 1 > {}^-5$
$$x > {}^-5 - 1$$
$$x > {}^-6$$
$$\therefore \quad {}^-6 < x < 4$$

$$^-6 < x < 4$$

```
|◄─┼──┼──┼──┼──○──┼──┼──┼──┼──┼──┼──┼──○──┼──┼──┼──┼──┼──┼──►|
 -10 -9 -8 -7 -6 -5 -4 -3 -2 -1  0  1  2  3  4  5  6  7  8  9 10
```

(22) $|10 - 2x| \geq 20$

$10 - 2x \geq 20$

$-2x \geq 20 - 10$

$-2x \geq 10$

$$\frac{-2x}{-2} \leq \frac{10}{-2}$$

$x \leq {}^-5$

OR $10 - 2x \leq {}^-20$

$-2x \leq {}^-20 - 10$

$-2x \leq {}^-30$

$$\frac{-2x}{-2} \geq \frac{{}^-30}{-2}$$

$x \geq 15$

(23) $|3x - 1| + 4 \leq 18$

$|3x - 1| + 4 - 4 \leq 18 - 4$

$|3x - 1| \leq 14$

$3x - 1 \leq 14$

$3x - 1 + 1 \leq 14 + 1$

$3x \leq 15$

$$\frac{3x}{3} \leq \frac{15}{3}$$

$x \leq 5$

OR $3x - 1 \geq {}^-14$

$3x - 1 + 1 \geq {}^-14 + 1$

$$3x \geq {}^-13$$

$$\frac{3x}{3} \geq \frac{{}^-13}{3}$$

$$x \geq \frac{{}^-13}{3}$$

$$\therefore \frac{{}^-13}{3} \leq x \leq 5$$

(24)
$$V = \frac{1}{3}\pi r^2 h$$
$$3V = \pi r^2 h$$
$$\pi r^2 h = 3V$$

$$\frac{\pi r^2 h}{\pi r^2} = \frac{3V}{\pi r^2}$$

$$h = \frac{3V}{\pi r^2}$$

(25)
Let the number be x
$$x + 13 = 30$$
$$x = 30 - 13$$
$$x = 17$$
$\therefore$ the number is 17

(26)
Let the number be x
$$x - 5 = 25$$
$$x = 25 + 5$$
$$x = 30$$
$\therefore$ The number is 30

(27)

$$\text{Let the number be } x$$
$$2(x - 5) = 100$$
$$2x - 10 = 100$$
$$2x = 100 + 10$$
$$2x = 110$$
$$\frac{2x}{2} = \frac{110}{2}$$
$$x = 55$$

$\therefore$ the number is 55

(28)

Let the 1st odd number be x
let the 2nd odd number be $x + 2$
$$x + (x + 2) = 24$$
$$x + x + 2 = 24$$
$$2x = 24 - 2$$
$$2x = 22$$
$$x = 11$$

$\therefore$ the 1st odd number is 11
the 2nd odd number is $11 + 2 = 13$

(29)

Let the 1st integer be x
Let the 2nd integer be $x + 1$
Let the 3rd integer be $x + 2$
$$x + (x + 1) + (x + 2) = 126$$
$$x + x + 1 + x + 2 = 126$$
$$3x + 3 = 126$$
$$3x = 123$$
$$\frac{3x}{3} = \frac{123}{3}$$
$$x = 41$$

$\therefore$ the 1st integer is 41
the 2nd integer is $41 + 1 = 42$
the 3rd integer is $41 + 2 = 43$

(30)

Let the 1st number be x

Then the 2nd number will be $33 - x$

$$\frac{x}{33 - x} = \frac{1}{2} \quad \text{cross multiply}$$

$$2x = 33 - x$$

$$2x + x = 33$$

$$3x = 33$$

$$\frac{3x}{3} = \frac{33}{3}$$

$$x = 11$$

∴   The 1st number is 11

The 2nd number is $33 - 11 = 22$

(31) Let the number of oranges given to a younger son be x

The elder son got $2x$

$$x + 2x = 33$$

$$3x = 33$$

$$\frac{3x}{3} = \frac{33}{3}$$

$$x = 11$$

∴   the younger son got 11 oranges

The elder son got $2 \times 11 = 22$ oranges

(32)

Let the number of all students in class be x

$\frac{2}{3}$ of all students are girls

$1 - \frac{2}{3} = \frac{1}{3}$ of all students are boys

$$\frac{1}{3}x = 20$$

$$x = 20 \times 3$$

$$x = 60$$

∴   the number of all students in class is 60

(33)

Let the number be x

$$\frac{1}{3} \, of \, x = \frac{1}{4} \, of \, x - 2$$

$$\frac{x}{3} = \frac{x}{4} - 2$$

Collect like terms together

$$\frac{x}{3} - \frac{x}{4} = {}^-2$$

$$\frac{4x - 3x}{12} = {}^-2$$

$$\frac{x}{12} = {}^-2$$

Multiply both sides by 12

$$x = {}^-24$$

∴ The number is ⁻24

(34)   Let the number of one-dollar bills be x

The number of five-dollar bills will be $44 - x$

$$1 * x + 5(44 - x) = 172$$

$$x + 220 - 5x = 172$$

$$x - 5x = 172 - 220$$

$$^-4x = {}^-48$$

$$\frac{^-4x}{^-4} = \frac{^-48}{^-4}$$

$$x = 12$$

∴   The number of one-dollar bills is 12

The number of five-dollar bills is $44 - 12 = 32$

(35)   C, $3x - 1 \le 8$

$$3x - 1 + 1 \le 8 + 1$$

$$3x \le 9$$

$$x \le 3$$

(36)   B, $3x - 1 > 8$

$3x - 1 + 1 > 8 + 1$

$3x > 9$

$x > 3$

Exercise 7B

Solve for x

(1) $5x + 17 = {}^-3$

(2) $x - 10 = 3x - 2$

(3) $4(x - 7) = 2(3x + 2)$

(4) $\dfrac{2}{3}x = \dfrac{4}{9}$

(5) $\dfrac{3x}{7} - \dfrac{x}{6} = \dfrac{121}{42}$

(6) $\dfrac{13x + 2}{3} = \dfrac{10x + 1}{2}$

(7) $\dfrac{1}{5}x - 3 = {}^-3x + \dfrac{1}{5}$

(8) $1.25x - 75 = 0.5x$

(9) $\sqrt{11x + 5} = 7$

(10) $3x^2 = 192$

(11) $1\dfrac{1}{4}x^3 = 80$

(12) $\dfrac{8}{x+2} - \dfrac{6}{x-2} = \dfrac{6}{x^2-4}$

(13) $\sqrt{x^2 - 15x + 75} = x + 5$

(14) $2x + 1 > 13$ and put your answer on a number line

(15) $\dfrac{^-2}{3}x \geq 8$ and put your answer on a number line

(16) $4x - 15 > 5x - 9$

(17) $^-x + 13 \leq x + 9$

(18) $\dfrac{x}{6} - 11 > 6$

(19) $\sqrt{3x + 49} \geq 10$

(20) $|x - 4| = 4$ and put your answer on a number line

(21) $|x + 3| < 7$ and put your answer on a number line

(22) $|10 - 5x| \geq 30$

(23) $|x - 2| + 4 \leq 20$

(24) Solve $A = \dfrac{1}{2}(a + b)\,h$ for b

(25) A number increased by 21 is 40. Find the number

(26) 105 less than a number is 95. Find the number

(27) Twice the difference between a number and 13 is 50. Find the number.

(28) Find two consecutive odd numbers whose sum is 44

(29) Find the three consecutive integers whose sum is 153

(30) The sum of two numbers is 60 and their quotient is $\dfrac{1}{3}$ Find the numbers.

(31) A mother had 57 oranges to give to her two sons. The elder son was given twice as much oranges as the younger son. How many oranges did the elder son get?

(32) Two-thirds of all students in a class are girls. There are

145

35 boys. Find the number of all students in the class.

(33) A two-third of a number is five less than a quarter of a number. Find the number.

(34) Joel has one-dollar bills and five-dollar bills in his bag. In his bag there are 22 bills together which has a value of 70. How many one-dollar bills and five-dollar bills are in John's bag?

(35) Which of the following equation has a right solution for the number line?

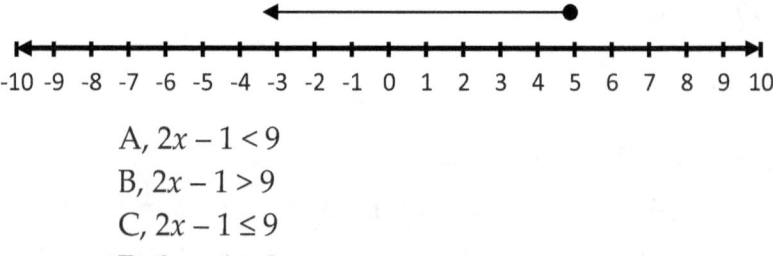

    A, $2x - 1 < 9$

    B, $2x - 1 > 9$

    C, $2x - 1 \leq 9$

    D, $2x - 1 \geq 9$

(36) Which of the following equation has a right solution for the number line?

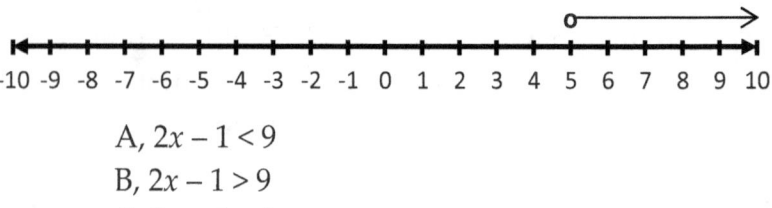

    A, $2x - 1 < 9$

    B, $2x - 1 > 9$

    C, $2x - 1 \leq 9$

    D, $2x - 1 \geq 9$

CHAPTER 8

POLYNOMIAL

Polynomial is an algebraic expression having 2 or more terms.

A <u>term</u> is a number(s) or variable(s) in an expression

<u>Variable</u> is a letter representing one or more numbers

<u>Algebraic expression</u> is an expression that has both variables and numbers for example $5x + xy + 2$

<u>Numerical expression</u> is an expression that has only numbers.

Degree of an expression $2x + 7$ is 1

Degree of an expression $3x^2 - x + 1$ is 2

Degree of an expression $5x^3 + x^2 - x + 3$ is 3

<u>Types of polynomial</u>

Monomial - $P(x) = 2x$, $P(x) = 3x^2$

Binomial - $P(x) = 5x^2 - 2$

Trinomial - $P(x) = 2x^2 + 4x + 3$

8-1 Add or Subtract polynomial
Examples 8.1

(a) $3ab^2 + 2c + 5ab^2 - c$

Collect like terms together

$3ab^2 + 5ab^2 + 2c - c$

$= 8ab^2 + c$

(b) $x^2 + 2x + y^2$

No like terms to add in an expression $x^2 + 2x + y^2$

8-2 Multiplying or Dividing polynomial
Examples 8.2

(a) $\qquad p(6a^2 * b) = 6a^2 bp$

(b) Simplify $\dfrac{18x^2}{6x} = \dfrac{18}{6} * \dfrac{x^2}{x}$

$$= 3x^{2-1}$$
$$= 3x$$

(c) $(2x^2 + 3x + 4) \div x + 1 =$

$$
\begin{array}{r}
2x + 1 \\
x+1\overline{\smash)2x^2 + 3x + 4} \\
-2x^2 + 2x \\
\hline
x + 4 \\
-x + 1 \\
\hline
3
\end{array}
$$

When dividing a polynomial,

$$\dfrac{\textbf{dividend}}{\textbf{divisor}} = \textbf{quotient} + \dfrac{\textbf{remainder}}{\textbf{divisor}}$$

$$\therefore (2x^2 + 3x + 4) \div x + 1 = 2x + 1 + \dfrac{3}{x+1}$$

(d) Evaluate $4x - 7$ when $x = 3$

$\qquad 4 * 3 - 7$

$\qquad 12 - 7 = 5$

8-3 Simplify or factorize using polynomial formulas
Examples 8.3

(a) From a formula $ab + ac = a(b + c)$

$\qquad 2x^2y + 2xy^2 = 2xy\,(x + y)$

(b) From a formula $(a + b)(c + d) = ac + ad + bc + bd$

$$(2x + y)(x + y) = 2x^2 + 2xy + xy + y^2$$
$$= 2x^2 + 3xy + y^2$$

(c) From a formula $(a + b)(a - b) = a^2 - b^2$

$$36x^2 - 9 = 9(4x^2 - 1)$$
$$= 9(2^2x^2 - 1)$$
$$= 9[(2x)^2 - 1^2]$$

From $a^2 - b^2 = (a + b)(a - b)$

$$9[(2x)^2 - 1^2] = 9(2x + 1)(2x - 1)$$

(d) From formula $(a + b)(a + b) = (a + b)^2 = a^2 + 2ab + b^2$

$$(2x + 3)^2 = (2x)^2 + 2*2x*3 + 3^2$$
$$= 4x^2 + 12x + 9$$

(e) From a formula $(a - b)(a - b) = (a - b)^2 = a^2 - 2ab + b^2$

$$(2 - x)^2 = 2^2 - 2 * 2 * x + x^2$$
$$= 4 - 4x + x^2$$

(f) From a formula

$$(a + b)(a + b)(a + b) = (a + b)^3 = a^3 + 3a^2b + 3ab^2 + b^3$$
$$(x + 2)^3 = x^3 + 3 * x^2 * 2 + 3 * x * 2^2 + 2^3$$
$$= x^3 + 6x^2 + 12x + 8$$

(g) From a formula

$$(a - b)(a - b)(a - b) = (a - b)^3 = a^3 - 3a^2b + 3ab^2 - b^3$$
$$\frac{1}{8}(2x - 2)^3 = \frac{1}{8}[2(x - 1)]^3$$

$$= \frac{1}{8} * 2^3(x - 1)^3$$

$$= \frac{1}{8}*8(x - 1)^3$$

DAVID M KASASA

$$= (x-1)^3$$
$$= x^3 - 3*x^2*1 + 3*x*1^2 - 1^3$$
$$= x^3 - 3x^2 + 3x - 1$$

(h) From a formula $(a+b)(a^2 - ab + b^2) = a^3 + b^3$
$$8x^3 + 27 = 2^3x^3 + 3^3$$
$$(2x)^3 + 3^3 = (2x+3)[(2x)^2 - 2x*3 + 3^2]$$
$$= (2x+3)(4x^2 - 6x + 9)$$

(i) From a formula $(a-b)(a^2 + ab + b^2) = a^3 - b^3$
$$(2x)^3 - \left(\frac{1}{2}\right)^3 = (2x-\frac{1}{2})[(2x)^2 + 2x*\frac{1}{2} + \left(\frac{1}{2}\right)^2]$$

$$= (2x-\frac{1}{2})(4x^2 + x + \frac{1}{4})$$

Other polynomial formulas are:
$$(a+b+c)^2 = a^2 + b^2 + c^2 + 2ab + 2ac + 2bc$$
$$(a-b)(a^3 + a^2b + ab^2 + b^3) = a^4 - b^4$$
$$(a-b)(a^{n-1} + a^{n-2}b + a^{n-3}b^2 + \ldots$$
$$\ldots + ab^{n-2} + b^{n-1}) = a^n - b^n$$
n is a positive integer (1, 2, 3, 4,.....)

$$(a+b)(a^{n-1} - a^{n-2}b + a^{n-3}b^2 - \ldots$$
$$\ldots - ab^{n-2} + b^{n-1}) = a^n + b^n$$
n is a positive odd integer (1, 3, 5, 7,.....)

The binomial theorem

$$(x+y)^n = x^n + a_{n-1}x^{n-1}y + a_{n-2}x^{n-2}y^2 + \ldots + y^n$$

Where coefficients a_i are called binomial coefficients and they form the rows of Pascal's Triangle.

			1			
		1		1		
	1		2		1	
1		3		3		1
1	4	6	4	1		
1	5	10	10	5	1	
1	6	15	20	15	6	1

$(x+y)^0 =$ ⟶ 1

$(x+y)^1 =$ ⟶ $1x + 1y$

$(x+y)^2 =$ ⟶ $1x^2 + 2xy + 1y^2$

$(x+y)^3 =$ ⟶ $1x^3 + 3x^2y + 3xy^2 + 1y^3$

1	4	6	4	1
1	5	10	10	5	1
1	6	15	20	15	6	1

This goes on and on and any number in the Pascal's triangle is equal to the sum of its above two numbers.

Examples 8.4

(a) Expand $(3p + 2)^2$

From $(x + y)^2 = x^2 + 2xy + y^2$ (Pascal's triangle above)

$$(3p + 2)^2 = (3p)^2 + 2 * 3p * 2 + 2^2$$
$$= 9p^2 + 12p + 4$$

(b) Find the ratio $x : y$ in the equation $\dfrac{x + y}{2} = \dfrac{x - y}{3}$

Solution

$$\frac{x + y}{2} = \frac{x - y}{3}$$

Cross multiply

$$3(x + y) = 2(x - y)$$
$$3x + 3y = 2x - 2y$$

Collect like terms together

$$3x - 2x = {}^-2y - 3y$$
$$x = {}^-5y$$

Divide both sides by y

$$\frac{x}{y} = \frac{{}^-5y}{y}$$

$$\frac{x}{y} = {}^-5$$

$\therefore$ the ratio x to y or $x : y = {}^-5$

(c) Decompose a rational expression of $\dfrac{2x + 8}{x^2 - 4}$ into partial fractions.

Solution

$$\frac{2x + 8}{x^2 - 4} = \frac{A}{x + 2} + \frac{B}{x - 2}$$

Since $x^2 - 4 = (x + 2)(x - 2)$

$$\frac{2x + 8}{(x + 2)(x - 2)} = \frac{A}{x + 2} + \frac{B}{x - 2}$$

Multiply each term by $(x + 2)(x - 2)$

$$\frac{2x+8}{(x+2)(x-2)}(x + 2)(x - 2) = \frac{A}{x+2}(x + 2)(x - 2) + \frac{B}{x-2}(x + 2)(x - 2)$$

$$2x + 8 = A(x - 2) + B(x + 2)$$

$$A(x - 2) + B(x + 2) = 2x + 8 \qquad (1)$$

When $x = 2$

$$A(2 - 2) + B(2 + 2) = 2(2) + 8$$

$$A(0) + B(4) = 4 + 8$$

$$4B = 12$$

$$\frac{4B}{4} = \frac{12}{4}$$

$$B = 3$$

From equation (1) when $x = {}^-2$

$$A(x - 2) + B(x + 2) = 2x + 8$$

$$A({}^-2 - 2) + B({}^-2 + 2) = 2({}^-2) + 8$$

$$A({}^-4) + B(0) = {}^-4 + 8$$

$${}^-4A = 4$$

$$\frac{{}^-4A}{{}^-4} = \frac{4}{{}^-4}$$

$$A = {}^-1$$

From $\dfrac{2x+8}{x^2-4} = \dfrac{A}{x+2} + \dfrac{B}{x-2}$

$\dfrac{2x+8}{x^2-4} = \dfrac{^-1}{x+2} + \dfrac{3}{x-2}$

$\dfrac{2x+8}{x^2-2} = \dfrac{3}{x-2} - \dfrac{1}{x+2}$

Exercise 8A

(1) Find the least common denominator LCD or LCM of

(i) $3, 5x,$ (ii) $x^2, y-1$ (iii) $\dfrac{1}{x-1}, \dfrac{1}{x+1}$

(2) Simplify $7ab^2 + 7c + 3ab^2 - 2c$

(3) Simplify $a(8a^2 * 3b)$

(4) Divide $(3x^2 + 3x - 1) \div x + 2$

(5) Simplify $6x^2y + 3xy^2$

(6) Expand $(x+y)(3x-y)$

(7) Simplify $x^3y - xy^3$

(8) Simplify $72x^2 - 8$

(9) Expand $(\frac{1}{2}x + 4)^2$

(10) Expand $(7-x)^2$

(11) Simplify $\dfrac{2x^3y^2}{4xy}$

(12) Simplify $\dfrac{4 - x^2}{2 + x}$

(13) Simplify $\dfrac{8x}{x^2 - 4} - \dfrac{4}{x + 2}$

(14) Simplify $\dfrac{x^2 - 9}{3} \div \dfrac{x + 3}{9}$

(15) Divide a polynomial $\dfrac{x^3 - x^2 + 5x - 3}{x - 2}$

(16) Evaluate $\dfrac{1}{3}x + 16$ when $x = 12$

(17) Evaluate $xyz - y - 4$ when $x = 1, y = 2$ and $z = 3$

(18) Evaluate $a^2 + 2a - 2b$ if $a = 2$ and $b = {}^-2$

(19) Evaluate $2x^2 - \dfrac{1}{2}y$ if $x = {}^-2$ and $y = 10$

(20) Simplify $(2x + 3y) - (3x - y)$

(21) Simplify $3x^2 + 4x - 2(x^2 + x)$

(22) Simplify $125x^3 + 64$

(23) Simplify $\dfrac{1}{8}x^3 - 27$

(24) Simplify $(2x + 1)^3 + (1 - 2x)^3 + x^2 - 3$

(25) Expand $\left(2x + y + \dfrac{z}{2}\right)^2$

(26) Simplify $\dfrac{2ab}{a^2 - b^2} - \dfrac{b}{a - b} + 4$

(27) Find the ratio $x : y$ in the equation $\dfrac{2x - y}{3} = \dfrac{x - y}{2}$

(28) Find the ratio $x : y$ in the equation $\dfrac{3x + y}{2} = \dfrac{x - y}{3}$

(29) Decompose a rational expression of $\dfrac{2x + 4}{x^2 - 1}$ into a partial fractions.

SOLUTIONS FOR EXERCISE 8A

1(i) $\qquad 3 * 5x = 15x$

(ii) $\qquad x^2 * (y - 1) = x^2(y - 1)$

(iii) $\qquad (x - 1)(x + 1)$

(2) $\qquad 7ab^2 + 7c + 3ab^2 - 2c$

Collect like terms together

$7ab^2 + 3ab^2 + 7c - 2c$

$$= 10ab^2 + 5c$$
$$= 5(2ab^2 + c)$$

(3) $\qquad a(8a^2 * 3b)$
$$= a(24a^2 b)$$
$$= 24a^3 b$$

(4) $\qquad (3x^2 + 3x - 1) \div x + 2 =$

$$
\begin{array}{r}
3x - 3 \\
x + 2 \overline{\smash{\big)}\, 3x^2 + 3x - 1} \\
\underline{-3x^2 + 6x} \\
{}^-3x - 1 \\
\underline{-{}^-3x - 6} \\
5
\end{array}
$$

When dividing a polynomial,

$$\frac{dividend}{divisor} = quotient + \frac{remainder}{divisor}$$

$$(3x^2 + 3x - 1) \div x + 2 = 3x - 3 + \frac{5}{x + 2}$$

$$(3x^2 + 3x - 1) \div x + 2 = 3(x - 1) + \frac{5}{x + 2}$$

(5) $$6x^2y + 3xy^2 = 3xy\,(2x + y)$$

(6)
$$(x + y)(3x - y) = 3x^2 - xy + 3xy - y^2$$
$$= 3x^2 + 2xy - y^2$$

(7)
$$x^3y - xy^3 = xy(x^2 - y^2)$$
From $a^2 - b^2 = (a + b)\,(a - b)$
$$xy(x^2 - y^2) = xy(x + y)\,(x - y)$$

(8)
$$72x^2 - 8 = 8(9x^2 - 1)$$
$$= 8(3^2x^2 - 1)$$
$$= 8[(3x)^2 - 1^2]$$
From $a^2 - b^2 = (a + b)\,(a - b)$
$$8[(3x)^2 - 1^2] = 8(3x + 1)(3x - 1)$$

(9) $$(\tfrac{1}{2}x + 4)^2$$

From $(a + b)^2 = a^2 + 2ab + b^2$
$$(\tfrac{1}{2}x + 4)^2 = \left(\tfrac{1}{2}x\right)^2 + 2*\tfrac{1}{2}x*4 + 4^2$$
$$= \frac{1}{4}x^2 + 4x + 16$$

(10) $$(7 - x)^2$$

From $(a - b)^2 = a^2 - 2ab + b^2$
$$(7 - x)^2 = 7^2 - 2 * 7 * x + x^2$$
$$= 49 - 14x + x^2$$

(11) $$\frac{2x^3y^2}{4xy} = \frac{x^2y^1}{2}$$

$$= \frac{1}{2}(x^2 y)$$

(12)
$$\frac{4-x^2}{2+x} = \frac{2^2-x^2}{2+x}$$

$$= \frac{(2+x)(2-x)}{2+x}$$

$$= 2 - x$$

(13)
$$\frac{8x}{x^2-4} - \frac{4}{x+2} = \frac{8x}{(x-2)(x+2)} - \frac{4}{x+2}$$

$$= \frac{8x - 4(x-2)}{(x-2)(x+2)}$$

$$= \frac{8x - 4x + 8}{(x-2)(x+2)}$$

$$= \frac{4x + 8}{(x-2)(x+2)}$$

$$= \frac{4(x+2)}{(x-2)(x+2)}$$

$$= \frac{4}{x-2}$$

(14)
$$\frac{x^2-9}{3} \div \frac{x+3}{9} = \frac{x^2-3^2}{3} \div \frac{x+3}{9}$$

$$= \frac{(x+3)(x-3)}{3} \div \frac{x+3}{9}$$

$$= \frac{(x+3)(x-3)}{3} \times \frac{9}{x+3}$$

$$= 3(x-3)$$

(15) $\dfrac{x^3 - x^2 + 5x - 3}{x - 2}$

$$
\begin{array}{r}
x^2 + x + 7 \\
x - 2 \overline{\smash{)}\, x^3 - x^2 + 5x - 3} \\
-\ x^3 - 2x^2 \\
\hline
x^2 + 5x \\
-\ x^2 - 2x \\
\hline
7x - 3 \\
-\ 7x - 14 \\
\hline
11
\end{array}
$$

∴ $\dfrac{x^3 - x^2 + 5x - 3}{x - 2} = x^2 + x + 7 + \dfrac{11}{x - 2}$

(16)
$$\frac{1}{3}x + 16 = \frac{1}{3} * 12 + 16$$
$$= 4 + 16$$
$$= 20$$

(17)
$$xyz - y - 4 = 1 * 2 * 3 - 2 - 4$$
$$= 6 - 6$$
$$= 0$$

(18)
$$a^2 + 2a - 2b = 2^2 + 2 * 2 - 2 * {}^-2$$
$$= 4 + 4 + 4$$
$$= 12$$

(19)
$$2x^2 - \frac{1}{2}y = 2 * {}^-2^2 - \frac{1}{2} * 10$$
$$= 8 - 5$$
$$= 3$$

(20) $(2x + 3y) - (3x - y) = 2x + 3y - 3x + y$
$$= 2x - 3x + 3y + y$$
$$= {}^-x + 4y$$

(21) $3x^2 + 4x - 2(x^2 + x) = 3x^2 + 4x - 2x^2 - 2x$
$$= 3x^2 - 2x^2 + 4x - 2x$$
$$= x^2 + 2x$$
$$= x(x + 2)$$

(22)
$$125x^3 + 64 = 5^3x^3 + 4^3$$
$$= (5x)^3 + 4^3$$
From $a^3 + b^3 = (a + b)(a^2 - ab + b^2)$
$$(5x)^3 + 4^3 = (5x + 4)[(5x)^2 - 5x * 4 + 4^2]$$
$$= (5x + 4)(25x^2 - 20x + 16)$$

(23)
$$\frac{1}{8}x^3 - 27 = \left(\frac{1}{2}\right)^3 x^3 - 3^3$$

$$= \left(\frac{1}{2}x\right)^3 - 3^3$$
From $a^3 - b^3 = (a - b)(a^2 + ab + b^2)$

$$\left(\frac{1}{2}x\right)^3 - 3^3 = (\frac{1}{2}x - 3)[(\frac{1}{2}x)^2 + \frac{1}{2}x * 3 + 3^3]$$

$$= (\frac{1}{2}x - 3)(\frac{1}{4}x^2 + \frac{3}{2}x + 9)$$

(24) $(2x + 1)^3 + (1 - 2x)^3 + x^2 - 3$

$$(2x)^3 + 3(2x)^2 1 + 3 * 2x(1)^2 + (1)^3 +$$
$$(1)^3 - 3(1)^2 2x + 3 * 1(2x)^2 - (2x)^3 + x^2 - 3$$
$$8x^3 + 12x^2 + 6x + 1 + 1 - 6x + 12x^2 - 8x^3 + x^2 - 3$$
$$8x^3 - 8x^3 + 12x^2 + 12x^2 + x^2 + 6x - 6x + 1 + 1 + -3$$
$$(12 + 12 + 1) x^2 + 1 + 1 - 3$$
$$25x^2 - 1$$
$$= 5^2 x^2 - 1^2$$
$$= (5x)^2 - 1^2$$
$$= (5x + 1)(5x - 1)$$

(25) $\left(2x + y + \dfrac{z}{2}\right)^2$

From $(a + b + c)^2 = a^2 + b^2 + c^2 + 2ab + 2ac + 2bc$

$= (2x)^2 + y^2 + \left(\dfrac{z}{2}\right)^2 + 2(2x)y + 2(2x)\left(\dfrac{z}{2}\right) + 2y\left(\dfrac{z}{2}\right)$

$= 4x^2 + y^2 + \dfrac{z^2}{4} + 4xy + 2xz + yz$

(26) $\dfrac{2ab}{a^2 - b^2} - \dfrac{b}{a - b} + 4$

$= \dfrac{2ab}{(a + b)(a - b)} - \dfrac{b}{a - b} + 4$

$= \dfrac{2ab - b(a + b) + 4(a + b)(a - b)}{(a + b)(a - b)}$

$= \dfrac{2ab - ab - b^2 + 4(a + b)(a - b)}{(a + b)(a - b)}$

$= \dfrac{ab - b^2 + 4(a + b)(a - b)}{(a + b)(a - b)}$

$$= \frac{b(a-b) + 4(a+b)(a-b)}{(a+b)(a-b)}$$

$$= \frac{(a-b)[b + 4(a+b)]}{(a+b)(a-b)}$$

$$= \frac{(a-b)(b + 4a + 4b)}{(a+b)(a-b)}$$

$$= \frac{(a-b)(4a + 5b)}{(a+b)(a-b)}$$

$$= \frac{4a + 5b}{a+b}$$

(27) $\quad \dfrac{2x - y}{3} = \dfrac{x - y}{2}$

Cross multiply

$2(2x - y) = 3(x - y)$

$4x - 2y = 3x - 3y$

Collect like terms together

$4x - 3x = 2y - 3y$

$x = {}^-y$

Divide both sides by y

$$\frac{x}{y} = \frac{{}^-y}{y}$$

$$\frac{x}{y} = {}^-1$$

∴ The ratio x to y or $x : y = {}^-1$

(28) $\quad \dfrac{3x + y}{2} = \dfrac{x - y}{3}$

Cross multiply

$$3(3x + y) = 2(x - y)$$
$$9x + 3y = 2x - 2y$$

Collect like terms together

$$9x - 2x = {}^-2y - 3y$$
$$7x = {}^-5y$$

Divide both sides by y

$$\frac{7x}{y} = \frac{{}^-5y}{y}$$

$$\frac{7x}{y} = {}^-5$$

Divide both sides by 7

$$\frac{7x}{7y} = \frac{{}^-5}{7}$$

$$\frac{x}{y} = \frac{{}^-5}{7}$$

$\therefore$ The ratio x to y or $x : y = {}^-5 : 7$

(29) $$\frac{2x + 4}{x^2 - 1} = \frac{A}{x + 1} + \frac{B}{x - 1}$$

From $x^2 - 1 = (x + 1)(x - 1)$

$$\frac{2x + 4}{(x + 1)(x - 1)} = \frac{A}{x + 1} + \frac{B}{x - 1}$$

Multiply each term by $(x + 1)(x - 1)$

$$\frac{2x+4}{(x+1)(x-1)}(x + 1)(x- 1) = \frac{A}{x+1}(x + 1)(x- 1) + \frac{B}{x-1}(x + 1)(x- 1)$$

$$2x + 4 = A(x- 1) + B(x + 1)$$
$$A(x- 1) + B(x + 1) = 2x + 4 \qquad\qquad (1)$$

When $x = 1$

$$A(1-1) + B(1+1) = 2(1) + 4$$

$$A(0) + B(2) = 2 + 4$$

$$2B = 6$$

$$\frac{2B}{2} = \frac{6}{2}$$

$$B = 3$$

From equation (1)

$$A(x-1) + B(x+1) = 2x + 4$$

When $x = {}^{-}1$

$$A({}^{-}1-1) + B({}^{-}1+1) = 2({}^{-}1) + 4$$

$$A({}^{-}2) + B(0) = {}^{-}2 + 4$$

$${}^{-}2A = 2$$

$$\frac{{}^{-}2A}{{}^{-}2} = \frac{2}{{}^{-}2}$$

$$A = {}^{-}1$$

From $\quad \dfrac{2x+4}{x^2-1} = \dfrac{A}{x+1} + \dfrac{B}{x-1}$

$$\frac{2x+4}{x^2-1} = \frac{{}^{-}1}{x+1} + \frac{3}{x-1}$$

$$\frac{2x+4}{x^2-1} = \frac{3}{x-1} - \frac{1}{x+1}$$

Exercise 8B

(1) Find the least common denominator LCD or LCM of

(i) $7, 3x,$ (ii) $x^2, x - y$ (iii) $\dfrac{1}{x-2}, \dfrac{1}{x+2}$

(2) Simplify $9ab^2 + 8ac + 6ab^2 - 3ac$

(3) Simplify $b(2a^2 * 3b)$

(4) Divide $(2x^2 + 7x + 5) \div x + 3$

(5) Simplify $12x^2y + 4xy^2$

(6) Expand $(2x + y)(x - 3y)$

(7) Simplify $x^3yz^2 - xw^2y^3$

(8) Simplify $5x^2 - 45$

(9) Expand $(\dfrac{1}{3}x + 6)^2$

(10) Expand $(11 - x)^2$

(11) Simplify $\dfrac{63x^3y^2}{9xy}$

(12) Simplify $\dfrac{16 - x^2}{4 + x}$

(13) Simplify $\dfrac{8x}{x^2 - 25} - \dfrac{4}{x + 5}$

(14) Simplify $\dfrac{x^2 - 36}{2} \div \dfrac{x + 6}{4}$

(15) Divide a polynomial $\dfrac{2x^3 - 9x^2 + 9x}{x - 3}$

(16) Divide a polynomial $\dfrac{x^4 + 2x^3 + 3x^2 + 4x + 1}{x + 2}$

(17) Evaluate $\dfrac{1}{13}x + 10$ when $x = 39$

(18) Evaluate $xyz - xy - z$ when $x=2$, $y=3$ and $z=4$

(19) Evaluate $x^2 + 2x - 2y$ if $x = 4$ and $y = {}^-3$

(20) Evaluate $x^2 - \dfrac{1}{7}y$ if $x = {}^-10$ and $y = 7$

(21) Simplify $({}^-3x + y) - (x - 2y)$

(22) Simplify $6x^2 - 2(x^2 + x)$

(23) Simplify $64x^3 + 216$

(24) Simplify $\dfrac{1}{2}x^3 - 4$

(25) Simplify $(1 + 3x)^3 + (1 - 3x)^3 - 5x^2 - 6$

(26) Expand $\left(\dfrac{x}{2} + 2y + z\right)^2$

(27) Simplify $\dfrac{2xy}{x^2 - y^2} - \dfrac{x}{x - y} + 2$

(28) Find the ratio $x : y$ in the equation $\dfrac{x - 3y}{2} = \dfrac{2x - y}{3}$

(29) Find the ratio $x : y$ in the equation $\dfrac{x + y}{2} = \dfrac{2x - 3y}{3}$

(30) Find the ratio $y : x$ in the equation $\dfrac{4xy - 6x}{14y^2 - 21y} = 1$

(31) Decompose a rational expression of $\dfrac{x + 9}{x^2 - 9}$ into a partial fractions.

(32) Decompose a rational expression of $\dfrac{3x-4}{x^2-16}$ into a partial fractions.

CHAPTER 9

QUADRATIC EQUATION

Quadratic equation or quadratic polynomial is an algebraic expression of degree 2 given in this form $P_{(x)} = ax^2 + bx + c$, where a, b, and c are integers and $a \neq 0$.

9-1 Factoring a quadratic polynomial $P_{(x)} = ax^2 + bx + c$

Find a pair or two factors of acx^2 whose sum is bx. If there are no factors whose sum is bx, then a polynomial cannot be factorized and it said to be <u>prime polynomial.</u>

Examples 9.1
Factorize the following:
(a) $m^2 + 5nm + 6n^2$

Solution

$$m^2 + 5nm + 6n^2$$
Multiply $6n^2$ and m^2

$$6n^2 m^2 \diagdown \begin{matrix} 2nm \\ 3nm \end{matrix}$$

a pair or 2 factors whose sum is $5nm$
$$m^2 + 2nm + 3nm + 6n^2$$
$$m(m + 2n) + 3n(m + 2n)$$
$$(m + 2n)(m + 3n)$$

(b) $\qquad 2x^2 + x + 1$

Solution

$2x^2 + x + 1$
Multiply $2x^2$ and 1

$2x^2$ ↗↘

No pair or 2 factors whose sum is x
Therefore $2x^2 + x + 1$ cannot be factorized,
it is a Prime Polynomial.

(c) $\qquad x^2 + 3x + 2$

Solution

$x^2 + 3x + 2$
Multiply x^2 and 2

$2x^2$ ↗x ↘$2x$

a pair or 2 factors whose sum is $3x$
$x^2 + x + 2x + 2$
$x(x + 1) + 2(x + 1)$
$(x + 1)(x + 2)$

9-2 Solving a Quadratic equation

There are 3 ways of how to solve a quadratic equation:
 (i) by Factorizing
 (ii) by Completing squares
 (iii) by quadratic formula

$$x = \frac{-b \pm \sqrt{b^2 - 4ac}}{2a}$$

Examples 9.2
Solve by Factorizing
a(i) $\qquad x^2 - x - 2 = 0$

Solution

$$x^2 - x - 2 = 0$$
Multiply x^2 and $^-2$

$$^-2x^2 \diagup\kern-0.6em\diagdown \begin{matrix} x \\ ^-2x \end{matrix}$$

a pair or 2 factors whose sum is ^-x
$$x^2 + x - 2x - 2 = 0$$
$$x(x + 1) - 2(x + 1) = 0$$
$$(x + 1)(x - 2) = 0$$
$$(x + 1) = 0$$
$$x + 1 - 1 = 0 - 1$$
$$x = {^-1}$$
When $\quad (x - 2) = 0$
$$x - 2 + 2 = 0 + 2$$
$$x = 2$$
$$\therefore\ x = \{^-1,\ 2\}$$

Solve by completing squares
(ii) $\qquad x^2 - x - 2 = 0$

Solution

$$x^2 - x - 2 = 0$$
Add the square of half the coefficient of x
$$x^2 - x + \left(\frac{1}{2}\right)^2 - 2 = 0 + \left(\frac{1}{2}\right)^2$$

$$x^2 - x + \frac{1}{4} - 2 = \frac{1}{4}$$

$$x^2 - x + \frac{1}{4} - 2 + 2 = \frac{1}{4} + 2$$

$$\left(x - \frac{1}{2}\right)^2 = \frac{1 + 8}{4}$$

$$\left(x - \frac{1}{2}\right)^2 = \frac{9}{4}$$

$$\sqrt{\left(x - \frac{1}{2}\right)^2} = \sqrt{\frac{9}{4}}$$

$$x - \frac{1}{2} = \pm \frac{3}{2}$$

Using the positive sign

$$x - \frac{1}{2} = \frac{3}{2}$$

$$x = \frac{3}{2} + \frac{1}{2}$$

$$x = \frac{4}{2}$$

$$x = 2$$

Using the negative sign

$$x - \frac{1}{2} = \frac{^-3}{2}$$

$$x = \frac{^-3}{2} + \frac{1}{2}$$

$$x = \frac{^-2}{2}$$

$$x = {}^-1$$
$$\therefore \quad x = \{{}^-1, 2\}$$

Solve by formula $x = \dfrac{-b \pm \sqrt{b^2 - 4ac}}{2a}$

Note the equation will not be solvable when <u>discriminant</u> (the value $b^2 - 4ac$) is a negative number.

(iii) $x^2 - x - 2 = 0$

Solution

$$x^2 - x - 2 = 0$$

Compare with $ax^2 + bx + c = 0$ and use a

formula $x = \dfrac{-b \pm \sqrt{b^2 - 4ac}}{2a}$

$a = 1, b = {}^-1, c = {}^-2$

$$x = \dfrac{-b \pm \sqrt{b^2 - 4ac}}{2a}$$

$$x = \dfrac{{}^-1 \pm \sqrt{{}^-1^2 - 4 * 1 * {}^-2}}{2 * 1}$$

$$x = \dfrac{1 \pm \sqrt{1 + 8}}{2}$$

$$x = \dfrac{1 \pm \sqrt{9}}{2}$$

$$x = \dfrac{1 \pm 3}{2}$$

Using the positive sign

$$x = \frac{1+3}{2}$$

$$x = \frac{4}{2}$$

$$x = 2$$

Using the negative sign
$$x = \frac{1-3}{2}$$

$$x = \frac{^-2}{2}$$

$$x = ^-1$$
$$\therefore \quad x = \{^-1, 2\}$$

b(i) $\qquad x^2 - 4x - 12 = 0$

Solution

$x^2 - 4x - 12 = 0$
Multiply x^2 and $^-12$

$^-12x^2 \diagdown \begin{array}{l} 2x \\ ^-6x \end{array}$

a pair or 2 factors whose sum is ^-4x
$x^2 + 2x - 6x - 12 = 0$
$x(x+2) - 6(x+2) = 0$
$(x+2)(x-6) = 0$
$\therefore (x+2) = 0$
$x + 2 - 2 = 0 - 2$
$x = ^- 2$
When $\quad (x-6) = 0$

$$x - 6 + 6 = 0 + 6$$
$$x = 6$$
$$\therefore \quad x = \{{}^-2, 6\}$$

(ii)   Using a formula $x^2 - 4x - 12 = 0$
Compare with $ax^2 + bx + c = 0$
$a = 1$, $b = {}^-4$, and $c = {}^-12$

$$x = \frac{-b \pm \sqrt{b^2 - 4ac}}{2a}$$

$$x = \frac{-{}^-4 \pm \sqrt{{}^-4^2 - 4*1*{}^-12}}{2*1}$$

$$x = \frac{4 \pm \sqrt{16 + 48}}{2}$$

$$x = \frac{4 \pm \sqrt{64}}{2}$$

$$x = \frac{4 \pm 8}{2}$$

Using the positive sign
$$x = \frac{4 + 8}{2}$$

$$x = \frac{12}{2}$$

$$x = 6$$

Using the negative sign

174

$$x = \frac{4 - 8}{2}$$

$$x = \frac{^-4}{2}$$

$$x = ^-2$$

$$\therefore \quad x = \{^-2, 6\}$$

(c)

$$(4 - 9x^2)(x - 1) = 0$$
$$(2^2 - 3^2x^2)(x - 1) = 0$$
$$[2^2 - (3x)^2](x - 1) = 0$$
$$(2 + 3x)(2 - 3x)(x - 1) = 0$$
$$2 + 3x = 0$$
$$3x = ^-2$$

$$\frac{3x}{3} = \frac{^-2}{3}$$

$$x = \frac{^-2}{3}$$

When $\quad 2 - 3x = 0$
$$- 3x = ^-2$$

$$\frac{^-3x}{^-3} = \frac{^-2}{^-3}$$

$$x = \frac{2}{3}$$

And when $\quad x - 1 = 0$
$$x = 1$$
$$\therefore x = \left\{\frac{^-2}{3}, \frac{2}{3}, 1\right\}$$

(d) $x^4 + 3x^2 + 2 = 0$ leave your answer in i form ($i^2 = {}^-1$)

Solution

Let $p = x^2$

$x^4 + 3x^2 + 2 = 0$

$p^2 + 3p + 2 = 0$

Multiply p^2 and 2

$$2p^2 \diagdown\diagup \begin{matrix} p \\ 2p \end{matrix}$$

a pair or 2 factors whose sum is $3p$

$p^2 + p + 2p + 2 = 0$

$p(p + 1) + 2(p + 1) = 0$

$(p + 1)(p + 2) = 0$

$p + 1 = 0$

$p = {}^-1$

But $p = x^2$

$x^2 = {}^-1$

$\sqrt{x^2} = \sqrt{{}^-1}$

But $i^2 = {}^-1$

$\sqrt{x^2} = \sqrt{i^2}$

$x = i$

When $p + 2 = 0$

$p = {}^-2$

$x^2 = {}^-2$

$\sqrt{x^2} = \sqrt{{}^-2}$

$\sqrt{x^2} = \sqrt{2 * {}^-1}$

But $i^2 = {}^-1$

$\sqrt{x^2} = \sqrt{2\,i^2}$

$x = i\sqrt{2}$

$\therefore$ $x = \{i, i\sqrt{2}\}$

(e) $$2x^3 - 5x^2 - 3x = 0$$

Solution

$$2x^3 - 5x^2 - 3x = 0$$
$$x(2x^2 - 5x - 3) = 0$$

Multiply $2x^2$ and $^-3$

$$^-6x^2 \diagup\!\!\!\diagdown \begin{array}{l} x \\ ^-6x \end{array}$$

a pair or 2 factors whose sum is ^-5x

$$x(2x^2 + x - 6x - 3) = 0$$
$$x[x(2x + 1) - 3(2x + 1)] = 0$$
$$x(2x + 1)(x - 3) = 0$$
$$x = 0$$

When $2x + 1 = 0$
$$2x = ^-1$$
$$x = \frac{^-1}{2}$$

And when $x - 3 = 0$
$$x = 3$$
$$\therefore \quad x = \{\tfrac{^-1}{2}, 0, 3\}$$

(f) $$\frac{x^4 - 7x^3 + 19x^2}{x^2} = 9$$

Solution

$$\frac{x^4 - 7x^3 + 19x^2}{x^2} = 9$$

$$\frac{x^2(x^2 - 7x + 19)}{x^2} = 9$$

177

$$x^2 - 7x + 19 = 9$$
$$x^2 - 7x + 19 - 9 = 0$$
$$x^2 - 7x + 10 = 0$$

Multiply x^2 and 10

$$10x^2 \diagup \overset{-2x}{\diagdown -5x}$$

a pair or 2 factors whose sum is $-7x$
$$x^2 - 2x - 5x + 10 = 0$$
$$x(x - 2) - 5(x - 2) = 0$$
$$(x - 2)(x - 5) = 0$$
$$x - 2 = 0$$
$$x = 2$$

When $\quad x - 5 = 0$
$$x = 5$$
$$\therefore \quad x = \{2, 5\}$$

Exercise 9A

Factorize and simplify completely

(1) $2m^2 + 7nm + 6n^2$

(2) $7x^2 - x + 1$

(3) $-2x^2 + 3x + 9$

(4), $\dfrac{x^2 - 7x + 12}{x^2 - 2x - 3}$

5(i) Simplify $\dfrac{x^2 + 2x - 15}{x^2 - 2x - 3}$

(ii) Find the values of x for which the polynomial or the rational expression is undefined

(iii) Find all values of x for which the polynomial or the rational expression equals to zero

Solve for x using one of the 3 ways in examples 9.2a.

(6) $x^2 + 6x + 8 = 0$

(7) $2x^2 + 5x + 10 = 7$

(8) $\dfrac{x + 16}{5} - \dfrac{x}{x - 2} = 2$

(9) $3x - \sqrt{3x} - 6 = 0$

(10) $\dfrac{1}{x - 2} = \dfrac{6}{x^2 - 4}$

(11) $x^3 - 5x^2 - 14x = 0$

(12) $\dfrac{x^4 - 8x^3 + 3x^2}{x^2} = 36$

(13) $(81 - 64x^2)(x - 3) = 0$

(14) Solve for x for values that $x > {}^-1$
$$\frac{(x - 7)(x^2 - 1)}{x + 1} = 0$$

(15) Solve for x for values that $x > 2$
$$\frac{(x^2 - x - 6)(2x - 5)}{x^2 - 4} = 0$$

(16) Solve for x and leave your answer in radical form
$$x^2 + 2x - 5 = 0$$

(17) Solve for x in the equation $x^2 - 2x + 5 = 0$ and leave

your answer in a form $p \pm qi$, where p and q are integers.

(18) Solve $x^4 - 5x^2 - 36 = 0$

(19) Use formula to solve $x^2 + 2x + 2 = 0$ and leave your answer in a form $p \pm qi$, where p and q are integers.

(20) The sum of two numbers is 7, and their product is 10 Find the numbers.

SOLUTIONS FOR EXERCISE 9A

(1) $2m^2 + 7nm + 6n^2$
Multiply $6n^2$ and $2m^2$

$12n^2m^2 \diagdown \begin{matrix} 4nm \\ 3nm \end{matrix}$

a pair or 2 factors whose sum is $7nm$
$2m^2 + 4nm + 3nm + 6n^2$
$2m(m + 2n) + 3n(m + 2n)$
$(m + 2n)(2m + 3n)$

(2) $7x^2 - x + 1$
Multiply $7x^2$ and 1

$7x^2 \diagdown$

No pair or 2 factors whose sum is ^-x
Therefore $7x^2 - x + 1$ cannot be factorized,
it is a Prime Polynomial.

(3) $^-2x^2 + 3x + 9$
Multiply $^-2x^2$ and 9

$^-18x^2 \diagdown \begin{matrix} 6x \\ ^-3x \end{matrix}$

a pair or 2 factors whose sum is $3x$
$^-2x^2 + 6x - 3x + 9$
$^-2x(x - 3) - 3(x - 3)$
$(x - 3)(^-2x - 3)$ OR $^-(x - 3)(2x + 3)$

(4)

$$\frac{x^2 - 7x + 12}{x^2 - 2x - 3}$$

For $x^2 - 7x + 12$

Multiply x^2 and 12

$12x^2 \diagdown \begin{matrix} -3x \\ -4x \end{matrix}$

a pair or 2 factors whose sum is $-7x$

$x^2 - 3x - 4x + 12$

$x(x - 3) - 4(x - 3)$

$(x - 3)(x - 4)$

For $x^2 - 2x - 3$

Multiply x^2 and -3

$-3x^2 \diagdown \begin{matrix} 1x \\ -3x \end{matrix}$

a pair or 2 factors whose sum is $-2x$

$x^2 + x - 3x - 3$

$x(x + 1) - 3(x + 1)$

$(x + 1)(x - 3)$

$$\frac{x^2 - 7x + 12}{x^2 - 2x - 3}$$

$$= \frac{(x - 3)(x - 4)}{(x + 1)(x - 3)}$$

$$= \frac{x - 4}{x + 1}$$

5(i)

$$\frac{x^2 + 2x - 15}{x^2 - 2x - 3}$$

For $x^2 + 2x - 15$

Multiply x^2 and $^-15$

$^-15x^2 \diagup\!\!\!\!\diagdown \begin{array}{l} ^-3x \\ 5x \end{array}$

a pair or 2 factors whose sum is $2x$

$x^2 - 3x + 5x - 15$

$x(x - 3) + 5(x - 3)$

$(x - 3)(x + 5)$

For $x^2 - 2x - 3$

Multiply x^2 and $^-3$

$^-3x^2 \diagup\!\!\!\!\diagdown \begin{array}{l} 1x \\ ^-3x \end{array}$

a pair or 2 factors whose sum is ^-2x

$x^2 + x - 3x - 3$

$x(x + 1) - 3(x + 1)$

$(x + 1)(x - 3)$

$$\frac{x^2 + 2x - 15}{x^2 - 2x - 3}$$

$$= \frac{(x - 3)(x + 5)}{(x + 1)(x - 3)}$$

$$= \frac{x + 5}{x + 1}$$

(ii)   When rational expression is undefined, the denominator is zero.

$\therefore$ $x^2 - 2x - 3 = 0$

$(x + 1)(x - 3) = 0$

$x + 1 = 0$

$$x = {}^-1$$

When $\quad x - 3 = 0$

$$x = 3$$

$$\therefore \ x = \{{}^-1, 3\}$$

(iii) $\qquad\qquad x^2 + 2x - 15 = 0$

$$(x - 3)(x + 5) = 0$$

$$x - 3 = 0$$

$$x = 3$$

When $\qquad x + 5 = 0$

$$x = {}^-5$$

$$\therefore \ x = \{{}^-5, 3\}$$

(6) $\qquad\qquad x^2 + 6x + 8 = 0$

Multiply x^2 and 8

$$8x^2 \Big\langle{}^{\textstyle 2x}_{\textstyle 4x}$$

a pair or 2 factors whose sum is $6x$

$$x^2 + 2x + 4x + 8 = 0$$

$$x(x + 2) + 4(x + 2) = 0$$

$$(x + 2)(x + 4) = 0$$

$$x + 2 = 0$$

$$x = {}^-2$$

When $\quad x + 4 = 0$

$$x = {}^-4$$

$$\therefore \ x = \{{}^-2, {}^-4\}$$

(7) $\qquad\qquad 2x^2 + 5x + 10 = 7$

$$2x^2 + 5x + 10 - 7 = 7 - 7$$

$$2x^2 + 5x + 3 = 0$$

Multiply $2x^2$ and 3

$$6x^2 \diagup\!\!\!\diagdown \begin{matrix} 2x \\ 3x \end{matrix}$$

a pair or 2 factors whose sum is $5x$

$2x^2 + 2x + 3x + 3 = 0$

$2x(x + 1) + 3(x + 1) = 0$

$(x + 1)(2x + 3) = 0$

$x + 1 = 0$

$x = {}^-1$

When $2x + 3 = 0$

$2x = {}^-3$

$x = \dfrac{{}^-3}{2}$

$\therefore \quad x = \{{}^-1, \dfrac{{}^-3}{2}\}$

(8)
$$\frac{x + 16}{5} - \frac{x}{x - 2} = 2$$

$$\frac{(x + 16)(x - 2) - 5x}{5(x - 2)} = 2$$

Multiply by $5(x - 2)$ on both sides

$(x + 16)(x - 2) - 5x = 2 * 5(x - 2)$

$(x + 16)(x - 2) - 5x = 10(x - 2)$

$x^2 - 2x + 16x - 32 - 5x = 10x - 20$

$x^2 - 2x + 16x - 5x - 10x = -20 + 32$

$x^2 - x = 12$

$x^2 - x - 12 = 0$

Multiply x^2 and $^-12$

$$-12x^2 \diagup\!\!\!\diagdown \begin{matrix} 3x \\ -4x \end{matrix}$$

a pair or 2 factors whose sum is ^-x

$$x^2 + 3x - 4x - 12 = 0$$
$$x(x + 3) - 4(x + 3) = 0$$
$$(x + 3)(x - 4) = 0$$
$$x + 3 = 0$$
$$x = {}^-3$$

When $\quad x - 4 = 0$
$$x = 4$$
$$\therefore \quad x = \{{}^-3, 4\}$$

(9) $\qquad 3x - \sqrt{3x} - 6 = 0$
$$3x - 3x - \sqrt{3x} - 6 = 0 - 3x$$
$${}^-\sqrt{3x} - 6 = {}^-3x$$
$${}^-\sqrt{3x} - 6 + 6 = {}^-3x + 6$$
$${}^-\sqrt{3x} = {}^-3x + 6$$
$$\sqrt{3x} = 3x - 6$$

Square both sides
$$(\sqrt{3x})^2 = (3x - 6)^2$$
$$3x = 9x^2 - 36x + 36$$
$$3x - 3x = 9x^2 - 36x - 3x + 36$$
$$0 = 9x^2 - 39x + 36$$
$$9x^2 - 39x + 36 = 0$$

Multiply $9x^2$ and 36

$$324x^2 \begin{cases} {}^-27x \\ {}^-12x \end{cases}$$

a pair or 2 factors whose sum is ${}^-39x$
$$9x^2 - 27x - 12x + 36 = 0$$
$$9x(x - 3) - 12(x - 3) = 0$$
$$(x - 3)(9x - 12) = 0$$
$$x - 3 = 0$$
$$x = 3$$

When $\quad 9x - 12 = 0$

$$9x = 12$$

$$x = \frac{12}{9}$$

$$x = \frac{4}{3}$$

Note this number when you check your answer $x = \frac{4}{3}$; it is not permissible.

Therefore $x = 3$ is the only answer.

OR

$$3x - \sqrt{3x} - 6 = 0$$

Let $n = \sqrt{3x}$

$$n^2 = 3x$$

$$n^2 - n - 6 = 0$$

Multiply n^2 and $^-6$

$$^-6n^2 \begin{array}{c} \nearrow 2n \\ \searrow ^-3n \end{array}$$

a pair or 2 factors whose sum is ^-n

$$n^2 + 2n - 3n - 6 = 0$$

$$n(n + 2) - 3(n + 2) = 0$$

$$(n + 2)(n - 3) = 0$$

$$n + 2 = 0$$

$$n = ^-2$$

But $n = \sqrt{3x}$

$$\sqrt{3x} = ^-2$$

Square both sides

$$3x = ^-2^2$$

$$3x = 4$$

$$x = \frac{4}{3}$$

When $n - 3 = 0$

$n = 3$

But $n = \sqrt{3x}$

$\sqrt{3x} = 3$

Square both sides

$3x = 3^2$

$3x = 9$

$x = 3$

Note this number when you check your answer $x = \dfrac{4}{3}$; it is not permissible.

Therefore $x = 3$ is the only answer.

(10)
$$\frac{1}{x - 2} = \frac{6}{x^2 - 4}$$

Cross multiply

$x^2 - 4 = 6(x - 2)$

$x^2 - 4 = 6x - 12$

$x^2 - 6x - 4 + 12 = 0$

$x^2 - 6x + 8 = 0$

Multiply x^2 and 8

$$8x^2 \begin{array}{l} \nearrow\ ^-2x \\ \searrow\ ^-4x \end{array}$$

a pair or 2 factors whose sum is ^-6x

$x^2 - 2x - 4x + 8 = 0$

$x(x - 2) - 4(x - 2) = 0$

$(x - 2)(x - 4) = 0$

$x - 2 = 0$

$x = 2$

When $\quad x - 4 = 0$

$x = 4$

Note this number when you check your

answer $x = 2$; it is not permissible

Therefore $x = 4$ is the only answer.

(11) $x^3 - 5x^2 - 14x = 0$

$x(x^2 - 5x - 14) = 0$

  Multiply x^2 and $^-14$

$^-14x^2 \diagup\!\!\!\diagdown \begin{matrix} 2x \\ -7x \end{matrix}$

a pair or 2 factors whose sum is ^-5x

$x(x^2 + 2x - 7x - 14) = 0$

$x[x(x + 2) - 7(x + 2)] = 0$

$x(x + 2)(x - 7) = 0$

$x = 0$

When $x + 2 = 0$

$x = ^-2$

And when $x - 7 = 0$

$x = 7$

∴ $x = \{^-2, 0, 7\}$

(12) $\dfrac{x^4 - 8x^3 + 3x^2}{x^2} = 36$

$\dfrac{x^2(x^2 - 8x + 3)}{x^2} = 36$

$x^2 - 8x + 3 = 36$

$x^2 - 8x + 3 - 36 = 0$

$x^2 - 8x - 33 = 0$

Multiply x^2 and $^-33$

$^-33x^2 \diagup\!\!\!\diagdown \begin{matrix} 3x \\ -11x \end{matrix}$

a pair or 2 factors whose sum is ^-8x

$$x^2 + 3x - 11x - 33 = 0$$

$$x(x + 3) - 11(x + 3) = 0$$

$$(x + 3)(x - 11) = 0$$

$$x + 3 = 0$$

$$x = {}^-3$$

When $\quad x - 11 = 0$

$$x = 11$$

$$\therefore \quad x = \{{}^-3, 11\}$$

(13) $\qquad (81 - 64x^2)(x - 3) = 0$

$$(9^2 - 8^2x^2)(x - 3) = 0$$

$$[9^2 - (8x)^2](x - 3) = 0$$

$$(9 + 8x)(9 - 8x)(x - 3) = 0$$

$$9 + 8x = 0$$

$$8x = {}^-9$$

$$\frac{8x}{8} = \frac{{}^-9}{8}$$

$$x = \frac{{}^-9}{8}$$

When $\quad 9 - 8x = 0$

$$-8x = {}^-9$$

$$\frac{{}^-8x}{{}^-8} = \frac{{}^-9}{{}^-8}$$

$$x = \frac{9}{8}$$

And when $\quad x - 3 = 0$

$$x = 3$$

$$\therefore \quad x = \left\{ \frac{-9}{8}, \ 1, \ \frac{9}{8} \right\}$$

(14)
$$\frac{(x - 7)(x^2 - 1)}{x + 1} = 0$$

$$\frac{(x - 7)(x + 1)(x - 1)}{x + 1} = 0$$

$$(x - 7)(x - 1) = 0$$
$$x - 7 = 0$$
$$x = 7$$

When $\quad x - 1 = 0$

$$x = 1$$

$\therefore \quad x = \{1, 7\}$

(15)
$$\frac{(x^2 - x - 6)(2x - 5)}{x^2 - 4} = 0$$

$$\frac{(x^2 + 2x - 3x - 6)(2x - 5)}{x^2 - 4} = 0$$

$$\frac{x(x + 2) - 3(x + 2)(2x - 5)}{x^2 - 2^2} = 0$$

$$\frac{(x + 2)(x - 3)(2x - 5)}{(x + 2)(x - 2)} = 0$$

$$\frac{(x - 3)(2x - 5)}{(x - 2)} = 0$$

$$(x - 3)(2x - 5) = 0$$
$$x - 3 = 0$$
$$x = 3$$

When $\quad 2x - 5 = 0$

$$2x = 5$$

$$x = \frac{5}{2}$$

$$\therefore \quad x = \left\{\frac{5}{2}, 3\right\}$$

(16) $\qquad x^2 + 2x - 5 = 0$

Compare with $ax^2 + bx + c = 0$ and use a

formula $\quad x = \dfrac{-b \pm \sqrt{b^2 - 4ac}}{2a}$

$a = 1, b = 2, c = {}^-5$

$$x = \frac{-b \pm \sqrt{b^2 - 4ac}}{2a}$$

$$x = \frac{-2 \pm \sqrt{2^2 - 4*1*{}^-5}}{2*1}$$

$$x = \frac{-2 \pm \sqrt{4 + 20}}{2}$$

$$x = \frac{-2 \pm \sqrt{24}}{2}$$

$$x = \frac{-2 \pm \sqrt{4 \times 6}}{2}$$

$$x = \frac{-2 \pm \sqrt{2^2 \times 6}}{2}$$

$$x = \frac{-2 \pm 2\sqrt{6}}{2}$$

$$x = \frac{2(-1 \pm \sqrt{6})}{2}$$

$$x = {}^-1 \pm \sqrt{6}$$

(17) $x^2 - 2x + 5 = 0$

Compare with $ax^2 + bx + c = 0$ and use a

formula $x = \dfrac{-b \pm \sqrt{b^2 - 4ac}}{2a}$

$a = 1, b = {}^-2, c = 5$

$$x = \frac{-b \pm \sqrt{b^2 - 4ac}}{2a}$$

$$x = \frac{-{}^-2 \pm \sqrt{{}^-2^2 - 4*1*5}}{2*1}$$

$$x = \frac{2 \pm \sqrt{4 - 20}}{2}$$

$$x = \frac{2 \pm \sqrt{{}^-16}}{2}$$

$$x = \frac{2 \pm \sqrt{{}^-1 \times 16}}{2}$$

But $i^2 = {}^-1$

$$x = \frac{2 \pm \sqrt{i^2 \times 4^2}}{2}$$

$$x = \frac{2 \pm \sqrt{(4i)^2}}{2}$$

$$x = \frac{2 \pm 4i}{2}$$

$$x = \frac{2(1 \pm 2i)}{2}$$

$$x = 1 \pm 2i$$

(18) $\qquad x^4 - 5x^2 - 36 = 0$

Let $n = x^2$

$(x^2)^2 - 5x^2 - 36 = 0$

$n^2 - 5n - 36 = 0$

Multiply n^2 and -36

$$-36n^2 \begin{cases} \nearrow 4n \\ \searrow -9n \end{cases}$$

a pair or 2 factors whose sum is $-5n$

$n^2 + 4n - 9n - 36 = 0$

$n(n + 4) - 9(n + 4) = 0$

$(n + 4)(n - 9) = 0$

$n + 4 = 0$

$n = -4$

But $n = x^2$

$x^2 = -4$

Square root both sides

$\sqrt{x^2} = \sqrt{-4}$

$x = \sqrt{(-1)4}$

But $i^2 = -1$

$x = \sqrt{i^2 2^2}$

$x = \pm 2i$

When $\quad n - 9 = 0$

$n = 9$

But $n = x^2$

194

$$x^2 = 9$$

Square root both sides

$$\sqrt{x^2} = \sqrt{9}$$

$$x = \sqrt{3^2}$$

$$x = \pm 3$$

$$\therefore \quad x = \{\pm 3, \pm 2i\}$$

(19) $\qquad$ $x^2 + 2x + 2 = 0$

Compare with $ax^2 + bx + c = 0$ and use a

formula $\quad x = \dfrac{-b \pm \sqrt{b^2 - 4ac}}{2a}$

$a = 1, b = 2, c = 2$

$$x = \frac{-b \pm \sqrt{b^2 - 4ac}}{2a}$$

$$x = \frac{-2 \pm \sqrt{2^2 - 4*1*2}}{2*1}$$

$$x = \frac{^-2 \pm \sqrt{4 - 8}}{2}$$

$$x = \frac{^-2 \pm \sqrt{^-4}}{2}$$

$$x = \frac{^-2 \pm \sqrt{(^-1)4}}{2}$$

But $i^2 = {}^-1$

$$x = \frac{^-2 \pm \sqrt{i^2 2^2}}{2}$$

$$x = \frac{^-2 \pm 2i}{2}$$

195

$$x = \frac{2(^-1 \pm i)}{2}$$

$$x = ^-1 \pm i$$

(20)

Let the 1st number be x
Then the 2nd number will be $7 - x$
Their product $x(7 - x) = 10$
$7x - x^2 = 10$
$^-x^2 + 7x - 10 = 0$
$x^2 - 7x + 10 = 0$
Multiply x^2 and 10

$$10x^2 \diagdown \begin{matrix} ^-2x \\ ^-5x \end{matrix}$$

a pair or 2 factors whose sum is ^-7x
$x^2 - 2x - 5x + 10 = 0$
$x(x - 2) - 5(x - 2) = 0$
$(x - 2)(x - 5) = 0$
$x - 2 = 0$
$x = 2$
When $x - 5 = 0$
$x = 5$
∴ when the 1st number is 2,
the 2nd number is $7 - 2 = 5$
when the 1st number is 5,
the 2nd number is $7 - 5 = 2$

Exercise 9B

Factorize and simplify completely
(1) $3a^2 + 10ab + 8b^2$
(2) $4x^2 - 2x + 1$
(3) $^-x^2 + 9x + 10$

(4) $\dfrac{2x^2 + x - 1}{2x^2 + 3x + 1}$

5(i) Simplify $\dfrac{x^2 - x - 12}{x^2 - 2x - 8}$

(ii) Find the values of x for which the polynomial or the rational expression is undefined

(iii) Find all values of x for which the polynomial or the rational expression equals to zero

Solve for x using one of the 3 ways in examples 9.2a.
(6) $x^2 + 8x + 12 = 0$
(7) $3x^2 + 13x + 6 = 2$

(8) $\dfrac{x + 4}{2} + \dfrac{x}{x - 2} = 2$

(9) $2x - \sqrt{9x} + 1 = 0$

(10) $\dfrac{1}{x - 6} = \dfrac{3}{x^2 - 36}$

(11) $x^3 + 6x^2 - 27x = 0$

(12) $\dfrac{x^4 - 6x^3 + 3x^2}{x^2} = 10$

(13) $\dfrac{x^3 + 2x^2 - 3x}{x} = 0$

(14) $(36 - 49x^2)(x - 4) = 0$

(15) Solve for x for values that $x > {}^-5$

$$\dfrac{(x - 8)(x^2 - 25)}{x + 5} = 0$$

(16) Solve for x for values that $x > {}^-5$

$$\dfrac{(x^2 - 17x + 72)(2x - 10)}{x^2 - 25} = 0$$

(17) Solve for x and leave your answer in radical form

$$x^2 + 4x - 3 = 0$$

(18) Solve for x and leave your answer in a form $p \pm qi$, where p and q are integers. $x^2 - 4x + 5 = 0$

(19) Solve $x^4 - 9x^2 + 20 = 0$

(20) Use formula to solve $x^2 - 2x + 2 = 0$ and leave your answer in a form $p \pm qi$, where p and q are integers.

(21) The sum of two numbers is 20, and their product is 99 Find the numbers.

CHAPTER 10

MATRICES

A matrix with 2 rows and 1 column$[a \quad b]$, is read 2×1 matrix or 2 by 1 matrix.

A matrix with 1 row and 2 columns$\begin{bmatrix} a \\ b \end{bmatrix}$, is read 1×2 matrix or 1 by 2 matrix.

A matrix with 2 rows and 2 columns$\begin{bmatrix} a & b \\ c & d \end{bmatrix}$, is read 2×2 matrix or 2 by 2 matrix.

10-1 Add or Subtract Matrices
Examples 10.1

$$\text{Let } p = \begin{bmatrix} {}^-1 & 2 \\ 3 & {}^-4 \end{bmatrix}$$

$$\text{and } q = \begin{bmatrix} 5 & 6 \\ 7 & 8 \end{bmatrix}$$

Find:

(i) -p

$$p = \begin{bmatrix} {}^-1 & 2 \\ 3 & {}^-4 \end{bmatrix}$$

$$\text{-}p = \begin{bmatrix} 1 & {}^-2 \\ {}^-3 & 4 \end{bmatrix}$$

-*p* is the Additive Inverse of matrix *p*.

(ii) p + q

$$\begin{bmatrix} {}^-1 & 2 \\ 3 & {}^-4 \end{bmatrix} + \begin{bmatrix} 5 & 6 \\ 7 & 8 \end{bmatrix} = \begin{bmatrix} 4 & 8 \\ 10 & 4 \end{bmatrix}$$

(iii) $3p - q$

$$3\begin{bmatrix} ^-1 & 2 \\ 3 & ^-4 \end{bmatrix} - \begin{bmatrix} 5 & 6 \\ 7 & 8 \end{bmatrix} = \begin{bmatrix} ^-3 & 6 \\ 9 & ^-12 \end{bmatrix} - \begin{bmatrix} 5 & 6 \\ 7 & 8 \end{bmatrix}$$

$$= \begin{bmatrix} ^-8 & 0 \\ 2 & ^-20 \end{bmatrix}$$

10-2 Multiply or Divide Matrixes
Examples 10.2

(i) $$[1 \quad 2]*\begin{bmatrix} 3 \\ 4 \end{bmatrix} = 1*3 + 2*4$$
$$= 3 + 8$$
$$= 11$$

11 is a *1* by *1* (*1×1*) Matrix

(ii) $$\begin{bmatrix} 3 \\ 2 \end{bmatrix}*[1 \quad 4] = \begin{bmatrix} 3*1 & 3*4 \\ 2*1 & 2*4 \end{bmatrix}$$

$$= \begin{bmatrix} 3 & 12 \\ 2 & 8 \end{bmatrix}$$

(iii) $$\begin{bmatrix} 1 & 0 \\ 2 & ^-1 \end{bmatrix}*\begin{bmatrix} 3 & 1 \\ 0 & 2 \end{bmatrix} = \begin{bmatrix} 1*3 + 0*0 & 1*1 + 0*2 \\ 2*3 + ^-1*0 & 2*1 + ^-1*2 \end{bmatrix}$$

$$= \begin{bmatrix} 3 & 1 \\ 6 & 0 \end{bmatrix}$$

(iv) Given $A = \begin{bmatrix} 6 & 2 \\ 3 & 0 \end{bmatrix}$,
Find $A \div 3$

$$\frac{A}{3} = \begin{bmatrix} \dfrac{6}{3} & \dfrac{2}{3} \\ \dfrac{3}{3} & \dfrac{0}{3} \end{bmatrix}$$

$$= \begin{bmatrix} 2 & \frac{2}{3} \\ 1 & 0 \end{bmatrix}$$

10-3 Determinant and the Inverse of a Matrix

If $P = \begin{bmatrix} a & b \\ c & d \end{bmatrix}$, In a 2 by 2 Matrix the determinant of P is given.

$$\det P = ad - bc$$

After finding the determinant of P, we can also find the inverse of P written as P^{-1}.

$$P = \begin{bmatrix} a & b \\ c & d \end{bmatrix},$$

$$P^{-1} = \frac{1}{\det P} \begin{bmatrix} d & ^{-}b \\ ^{-}c & a \end{bmatrix}$$

Note the changes of a and d in the formula
also the sign changes for b and c

$$P * P^{-1} = \begin{bmatrix} 1 & 0 \\ 0 & 1 \end{bmatrix} = I$$

It is called the Identity Matrix (I)

Examples 10.3

$$\text{Given } A = \begin{bmatrix} 1 & ^{-}2 \\ 3 & ^{-}1 \end{bmatrix},$$

(i)   Find $\det A = 1*^{-}1 - 3*^{-}2$

$$= ^{-}1 - ^{-}6$$

$$= ^{-}1 + 6$$

$$\det A = 5$$

(ii) Find the Inverse of A

$$A^{-1} = \frac{1}{5} \begin{bmatrix} ^{-}1 & 2 \\ ^{-}3 & 1 \end{bmatrix}$$

$$= \begin{bmatrix} \dfrac{^-1}{5} & \dfrac{2}{5} \\ \dfrac{^-3}{5} & \dfrac{1}{5} \end{bmatrix}$$

(iii)   Show that $A*A^{-1}$ is the Identity Matrix (I)

$$A*A^{-1} = \begin{bmatrix} 1 & ^-2 \\ 3 & ^-1 \end{bmatrix} * \begin{bmatrix} \dfrac{^-1}{5} & \dfrac{2}{5} \\ \dfrac{^-3}{5} & \dfrac{1}{5} \end{bmatrix}$$

$$= \begin{bmatrix} \dfrac{^-1}{5} + \dfrac{6}{5} & \dfrac{2}{5} + \dfrac{^-2}{5} \\ \dfrac{^-3}{5} + \dfrac{3}{5} & \dfrac{6}{5} + \dfrac{^-1}{5} \end{bmatrix}$$

$$= \begin{bmatrix} \dfrac{5}{5} & \dfrac{0}{5} \\ \dfrac{0}{5} & \dfrac{5}{5} \end{bmatrix}$$

$$= \begin{bmatrix} 1 & 0 \\ 0 & 1 \end{bmatrix}$$

$\therefore$ $A*A^{-1}$ is Identity Matrix (I) $= \begin{bmatrix} 1 & 0 \\ 0 & 1 \end{bmatrix}$

Exercise 10A

(1)   Let $p = \begin{bmatrix} 4 & 7 \\ 2 & -5 \end{bmatrix}$

and $q = \begin{bmatrix} -1 & 11 \\ 4 & -6 \end{bmatrix}$

Find:

(a) Additive inverse of matrices p and q

(b) $p + q$

(c) $2p - q$

(2) $[13 \quad {}^-2] * \begin{bmatrix} 2 \\ 9 \end{bmatrix} =$

(3) $\begin{bmatrix} 6 \\ 2 \end{bmatrix} * \begin{bmatrix} \frac{1}{2} & 4 \end{bmatrix} =$

(4) $\begin{bmatrix} 3 & 0 \\ 6 & -3 \end{bmatrix} * \begin{bmatrix} 3 & 1 \\ 0 & 2 \end{bmatrix} =$

(5) $\begin{bmatrix} 2 & -1 \\ 3 & 4 \end{bmatrix} * \begin{bmatrix} 2 & 1 & 5 \\ 3 & 4 & 6 \end{bmatrix} =$

(6) Given $R = \begin{bmatrix} 4 & 2 \\ 3 & 0 \end{bmatrix}$, Find $R \div 4$

(7) Given $P = \begin{bmatrix} 13 & -3 \\ 2 & -1 \end{bmatrix}$

(a) Find det P

(b) Find the Inverse of P

(c) Show that $P * P^{-1}$ is the Identity Matrix (I)

(8) Given $q = \begin{bmatrix} -7 & 3 \\ 1 & -3 \end{bmatrix}$

(a) Find det q

(b) Find the Inverse of q

(c) Show that $q * q^{-1}$ is the Identity Matrix (I)

SOLUTIONS FOR EXERCISE 10A

(1)
$$p = \begin{bmatrix} 4 & 7 \\ 2 & -5 \end{bmatrix}$$

$$\text{and } q = \begin{bmatrix} -1 & 11 \\ 4 & -6 \end{bmatrix}$$

(a) Additive inverse of matrices p and q

$$p = \begin{bmatrix} 4 & 7 \\ 2 & -5 \end{bmatrix}$$

$$-p = \begin{bmatrix} -4 & -7 \\ -2 & 5 \end{bmatrix}$$

$$q = \begin{bmatrix} -1 & 11 \\ 4 & -6 \end{bmatrix}$$

$$-q = \begin{bmatrix} 1 & -11 \\ -4 & 6 \end{bmatrix}$$

(b)   p + q

$$\begin{bmatrix} 4 & 7 \\ 2 & -5 \end{bmatrix} + \begin{bmatrix} -1 & 11 \\ 4 & -6 \end{bmatrix} = \begin{bmatrix} 3 & 18 \\ 6 & -11 \end{bmatrix}$$

(C)   2p − q

$$2\begin{bmatrix} 4 & 7 \\ 2 & -5 \end{bmatrix} - \begin{bmatrix} -1 & 11 \\ 4 & -6 \end{bmatrix} = \begin{bmatrix} 8 & 14 \\ 4 & -10 \end{bmatrix} - \begin{bmatrix} -1 & 11 \\ 4 & -6 \end{bmatrix}$$

$$= \begin{bmatrix} 9 & 3 \\ 0 & -4 \end{bmatrix}$$

(2)
$$[13 \quad -2] * \begin{bmatrix} 2 \\ 9 \end{bmatrix} = 13*2 + -2*9$$
$$= 26 - 18 = 8$$

(3)
$$\begin{bmatrix} 6 \\ 2 \end{bmatrix} * \begin{bmatrix} \frac{1}{2} & 4 \end{bmatrix} = \begin{bmatrix} 6 * \dfrac{1}{2} & 6 * 4 \\ 2 * \dfrac{1}{2} & 2 * 4 \end{bmatrix}$$

$$= \begin{bmatrix} 3 & 24 \\ 1 & 8 \end{bmatrix}$$

(4)
$$\begin{bmatrix} 3 & 0 \\ 6 & {}^-3 \end{bmatrix} * \begin{bmatrix} 3 & 1 \\ 0 & 2 \end{bmatrix} = \begin{bmatrix} 3*3 + 0*0 & 3*1 + 0*2 \\ 6*3 + {}^-3*0 & 6*1 + {}^-3*2 \end{bmatrix}$$

$$= \begin{bmatrix} 9 & 3 \\ 18 & 0 \end{bmatrix}$$

$$= 3 \begin{bmatrix} 3 & 1 \\ 6 & 0 \end{bmatrix}$$

(5)
$$\begin{bmatrix} 2 & {}^-1 \\ 3 & 4 \end{bmatrix} * \begin{bmatrix} 2 & 1 & 5 \\ 3 & 4 & 6 \end{bmatrix}$$

$$= \begin{bmatrix} 2*2 + {}^-1*3 & 2*1 + {}^-1*4 & 2*5 + {}^-1*6 \\ 3*2 + 4*3 & 3*1 + 4*4 & 3*5 + 4*6 \end{bmatrix}$$

$$= \begin{bmatrix} 4 + {}^-3 & 2 + {}^-4 & 10 + {}^-6 \\ 6 + 12 & 3 + 16 & 15 + 24 \end{bmatrix}$$

$$= \begin{bmatrix} 4 - 3 & 2 - 4 & 10 - 6 \\ 6 + 12 & 3 + 16 & 15 + 24 \end{bmatrix}$$

$$= \begin{bmatrix} 1 & {}^-2 & 4 \\ 18 & 19 & 39 \end{bmatrix}$$

(6)
$$R = \begin{bmatrix} 4 & 2 \\ 3 & 0 \end{bmatrix}$$

$$R \div 4 = \frac{R}{4}$$

$$= \begin{bmatrix} \dfrac{4}{4} & \dfrac{2}{4} \\ \dfrac{3}{4} & \dfrac{0}{4} \end{bmatrix}$$

$$= \begin{bmatrix} 1 & \dfrac{1}{2} \\ \dfrac{3}{4} & 0 \end{bmatrix}$$

7(a) $P = \begin{bmatrix} 13 & ^-3 \\ 2 & ^-1 \end{bmatrix}$

$$\det P = 13*^-1 - 2*^-3$$
$$= {}^-13 - {}^-6$$
$$= {}^-13 + 6$$
$$\det P = {}^-7$$

(b) $P^{-1} = \dfrac{1}{^-7} \begin{bmatrix} ^-1 & 3 \\ ^-2 & 13 \end{bmatrix}$

$$= \begin{bmatrix} \dfrac{1}{7} & \dfrac{3}{^-7} \\ \dfrac{^-2}{^-7} & \dfrac{13}{^-7} \end{bmatrix}$$

$$= \begin{bmatrix} \dfrac{1}{7} & \dfrac{^-3}{7} \\ \dfrac{2}{7} & \dfrac{^-13}{7} \end{bmatrix}$$

(c) $P*P^{-1} = \begin{bmatrix} 13 & ^-3 \\ 2 & ^-1 \end{bmatrix} * \begin{bmatrix} \dfrac{1}{7} & \dfrac{^-3}{7} \\ \dfrac{2}{7} & \dfrac{^-13}{7} \end{bmatrix}$

$$= \begin{bmatrix} \dfrac{13}{7} + \dfrac{^-6}{7} & \dfrac{^-39}{7} + \dfrac{39}{7} \\[4mm] \dfrac{2}{7} + \dfrac{^-2}{7} & \dfrac{^-6}{7} + \dfrac{13}{7} \end{bmatrix}$$

$$= \begin{bmatrix} \dfrac{7}{7} & \dfrac{0}{7} \\[4mm] \dfrac{0}{7} & \dfrac{7}{7} \end{bmatrix}$$

$$= \begin{bmatrix} 1 & 0 \\ 0 & 1 \end{bmatrix}$$

$\therefore P*P^{-1}$ is Identity Matrix (I) $= \begin{bmatrix} 1 & 0 \\ 0 & 1 \end{bmatrix}$

8(a) $q = \begin{bmatrix} ^-7 & 3 \\ 1 & ^-3 \end{bmatrix}$

$\det q = {}^-7*{}^-3 - 1*3$

$\quad\quad = 21 - 3$

$\det q = 18$

(b) $q^{-1} = \dfrac{1}{18} \begin{bmatrix} ^-3 & ^-3 \\ ^-1 & ^-7 \end{bmatrix}$

$$= \begin{bmatrix} \dfrac{^-3}{18} & \dfrac{^-3}{18} \\[4mm] \dfrac{^-1}{18} & \dfrac{^-7}{18} \end{bmatrix}$$

$$= \begin{bmatrix} \dfrac{^-1}{6} & -\dfrac{1}{6} \\[4mm] \dfrac{^-1}{18} & \dfrac{^-7}{18} \end{bmatrix}$$

(c)

$$q*q^{-1} = \begin{bmatrix} -7 & 3 \\ 1 & -3 \end{bmatrix} * \begin{bmatrix} \dfrac{-1}{6} & \dfrac{-1}{6} \\ \dfrac{-1}{18} & \dfrac{-7}{18} \end{bmatrix}$$

$$= \begin{bmatrix} \dfrac{7}{6}+\dfrac{-3}{18} & \dfrac{7}{6}+\dfrac{-21}{18} \\ \dfrac{-1}{6}+\dfrac{3}{18} & \dfrac{-1}{6}+\dfrac{21}{18} \end{bmatrix}$$

$$= \begin{bmatrix} \dfrac{18}{18} & \dfrac{0}{18} \\ \dfrac{0}{18} & \dfrac{18}{18} \end{bmatrix}$$

$$= \begin{bmatrix} 1 & 0 \\ 0 & 1 \end{bmatrix}$$

$\therefore\ q*q^{-1}$ is Identity Matrix (I) $= \begin{bmatrix} 1 & 0 \\ 0 & 1 \end{bmatrix}$

Exercise 10B

(1) Let $p = \begin{bmatrix} 6 & 7 \\ 4 & -3 \end{bmatrix}$

and $q = \begin{bmatrix} -3 & 13 \\ 8 & -4 \end{bmatrix}$

Find:

(a) Additive inverse of matrices p and q

(b) p + q

(c) 2p – q

(2) $[11 \quad -1] * \begin{bmatrix} 3 \\ 4 \end{bmatrix} =$

(3) $\begin{bmatrix} 4 \\ 6 \end{bmatrix} * \begin{bmatrix} \frac{1}{2} & 2 \end{bmatrix} =$

(4) $\begin{bmatrix} 5 & 7 \\ 0 & -1 \end{bmatrix} * \begin{bmatrix} 2 & 1 \\ 1 & 2 \end{bmatrix} =$

(5) $\begin{bmatrix} 3 & -2 \\ 4 & 0 \end{bmatrix} * \begin{bmatrix} 1 & 3 & 4 \\ 5 & 6 & 5 \end{bmatrix} =$

(6) Given $R = \begin{bmatrix} 16 & 22 \\ 1 & 0 \end{bmatrix}$, Find $R \div 2$

(7) Find a, b, c and d:

(i) $\frac{1}{3} \begin{bmatrix} a & b \\ c & d \end{bmatrix} = \begin{bmatrix} 12 & 1 \\ 5 & 3 \end{bmatrix}$

(ii) $\begin{bmatrix} 2 \\ 4 \end{bmatrix} * [a \quad b] = \begin{bmatrix} 10 & c \\ d & 16 \end{bmatrix}$

(8) Given $P = \begin{bmatrix} 7 & 3 \\ -2 & 2 \end{bmatrix}$

(a) Find det P

(b) Find the Inverse of P

(c) Show that $P*P^{-1}$ is the Identity Matrix (I)

(9) Given q = $\begin{bmatrix} 4 & 6 \\ 1 & 2 \end{bmatrix}$

(a) Find det q

(b) Find the Inverse of q

(c) Show that $q*q^{-1}$ is the Identity Matrix (I)

CHAPTER 11

SOLVING SYSTEMS OR SIMULTANEOUS EQUATION

In systems there are two or more variables (x,y) in two or more given equations.

There are three methods of solving simultaneous equations:

(a) by elimination
(b) by substitution
(c) by matrix

11-1 Solving by Elimination
Examples 11.1

(a) Solve for x and y

$$2x + y = 3$$
$$x - 3y = 5$$

Solution

$$2x + y = 3 \qquad \text{(i)}$$
$$x - 3y = 5 \qquad \text{(ii)}$$

Multiply equation *(i)* by 3

$$3(2x + y = 3)$$
$$6x + 3y = 9 \qquad \text{(iii)}$$

Add equation *(ii)* and *(iii)*

$$\begin{array}{rll} x - 3y &= 5 & \text{(ii)} \\ + \ 6x + 3y &= 9 & \text{(iii)} \\ \hline 7x &= 14 & \end{array}$$

Divide both sides by 7

$$\frac{7x}{7} = \frac{14^2}{7}$$

211

$$x = 2$$

From equation *(i)*

$$2x + y = 3$$
$$2*2 + y = 3$$
$$4 + y = 3$$
$$y = 3 - 4$$
$$y = {}^-1$$

(b) Solve for x and y

$$x + y = 3$$
$$3x - 2y = {}^-1$$

Solution

$$x + y = 3 \qquad \text{(i)}$$
$$3x - 2y = {}^-1 \qquad \text{(ii)}$$

Multiply equation *(i)* by 2

$$2(x + y = 3)$$
$$2x + 2y = 6 \qquad \text{(iii)}$$

Add equation (ii) and (iii)

$$3x - 2y = {}^-1 \qquad \text{(ii)}$$
$$+ \ 2x + 2y = 6 \qquad \text{(iii)}$$
_____
$$5x \qquad = 5$$

Divide both sides by 5

$$\frac{\cancel{5}x}{\cancel{5}} = \frac{\cancel{5}^1}{\cancel{5}}$$
$$x = 1$$

From equation *(i)*

$$x + y = 3$$
$$1 + y = 3$$
$$y = 3 - 1$$
$$y = 2$$

11-2 Solving by Substitution

Will be using different method on the above **Examples *11.1***
expecting same results.

Examples 11.2

(a)   Solve $2x + y = 3$

   $x - 3y = 5$

Solution

$$2x + y = 3 \qquad \text{(i)}$$
$$x - 3y = 5 \qquad \text{(ii)}$$

From *(ii)* add *3y* on both sides

$$x - 3y + 3y = 5 + 3y$$
$$x = 5 + 3y \qquad \text{(iii)}$$

Substitute x or equation *(iii)* into equation *(i)*

$$2x + y = 3 \qquad \text{(i)}$$
$$2(5 + 3y) + y = 3$$
$$10 + 6y + y = 3$$
$$10 + 7y = 3$$

Subtract *10* on both sides

$$10 - 10 + 7y = 3 - 10$$
$$7y = {}^-7$$

Divide both sides by *7*

$$\frac{7y}{7} = \frac{{}^-7}{7}$$

$$y = {}^-1$$

Substitute $y = {}^-1$ into equation (i)

$$2x + y = 3 \qquad \text{(i)}$$
$$2x + {}^-1 = 3$$
$$2x - 1 = 3$$
$$2x = 3 + 1$$
$$2x = 4$$

$$\frac{\cancel{2}x}{\cancel{2}} = \frac{4^2}{\cancel{2}}$$

$$x = 2$$

(b)   Solve for x and y

$$x + y = 3$$
$$3x - 2y = {}^-1$$

Solution

$$x + y = 3 \qquad\qquad\qquad \text{(i)}$$
$$3x - 2y = {}^-1 \qquad\qquad \text{(ii)}$$

From equation *(i)*

$$x + y = 3$$
$$x = 3 - y \qquad\qquad \text{(iii)}$$

Substitute x or equation *(iii)* into equation *(ii)*

$$3x - 2y = {}^-1 \qquad\qquad \text{(ii)}$$
$$3(3 - y) - 2y = {}^-1$$
$$9 - 3y - 2y = {}^-1$$
$$9 - 5y = {}^-1$$

Subtract 9 on both sides

$$9 - 9 - 5y = {}^-1 - 9$$
$${}^-5y = {}^-10$$

Divide by 5 on both sides

$$\frac{\cancel{{}^-5}y}{\cancel{{}^-5}} = \frac{\cancel{{}^-10}^2}{\cancel{{}^-5}}$$

$$y = 2$$

Substitute $y = 2$ into equation *(i)*

$$x + y = 3$$
$$x + 2 = 3$$
$$x = 3 - 2$$
$$x = 1$$

11-3 Solving by Matrix

Will be using different method on the above **Examples** *11.1* expecting same results.

Examples 11.3

(a)   Solve $2x + y = 3$

$\qquad\qquad x - 3y = 5$

Solution

$$2x + y = 3$$
$$x - 3y = 5$$

In matrix it becomes

$$\begin{bmatrix} 2 & 1 \\ 1 & ^-3 \end{bmatrix} \begin{bmatrix} x \\ y \end{bmatrix} = \begin{bmatrix} 3 \\ 5 \end{bmatrix} \qquad\qquad (i)$$

The inverse $= \dfrac{1}{(2*^-3)-(1*1)} \begin{bmatrix} ^-3 & ^-1 \\ ^-1 & 2 \end{bmatrix}$

$$= \dfrac{1}{^-6-1} \begin{bmatrix} ^-3 & ^-1 \\ ^-1 & 2 \end{bmatrix}$$

$$= \dfrac{1}{^-7} \begin{bmatrix} ^-3 & ^-1 \\ ^-1 & 2 \end{bmatrix}$$

$$= -\dfrac{1}{7} \begin{bmatrix} ^-3 & ^-1 \\ ^-1 & 2 \end{bmatrix}$$

$$= \begin{bmatrix} \dfrac{3}{7} & \dfrac{1}{7} \\ \dfrac{1}{7} & \dfrac{^-2}{7} \end{bmatrix}$$

Multiply the inverse on both sides of equation *(i)*

$$\begin{bmatrix} \dfrac{3}{7} & \dfrac{1}{7} \\ \dfrac{1}{7} & \dfrac{^-2}{7} \end{bmatrix} \begin{bmatrix} 2 & 1 \\ 1 & ^-3 \end{bmatrix} \begin{bmatrix} x \\ y \end{bmatrix} = \begin{bmatrix} \dfrac{3}{7} & \dfrac{1}{7} \\ \dfrac{1}{7} & \dfrac{^-2}{7} \end{bmatrix} \begin{bmatrix} 3 \\ 5 \end{bmatrix}$$

215

$$\begin{bmatrix} \frac{3}{7}*2 + \frac{1}{7}*1 & \frac{3}{7}*1 + \frac{1}{7}*^-3 \\ \frac{1}{7}*2 + \frac{^-2}{7}*1 & \frac{1}{7}*1 + \frac{^-2}{7}*^-3 \end{bmatrix} \begin{bmatrix} x \\ y \end{bmatrix} = \begin{bmatrix} \frac{3}{7}*3 + \frac{1}{7}*5 \\ \frac{1}{7}*3 + \frac{^-2}{7}*5 \end{bmatrix}$$

$$\begin{bmatrix} \frac{6}{7} + \frac{1}{7} & \frac{3}{7} - \frac{3}{7} \\ \frac{2}{7} - \frac{2}{7} & \frac{1}{7} + \frac{6}{7} \end{bmatrix} \begin{bmatrix} x \\ y \end{bmatrix} = \begin{bmatrix} \frac{9}{7} + \frac{5}{7} \\ \frac{3}{7} - \frac{10}{7} \end{bmatrix}$$

$$\begin{bmatrix} \frac{7}{7} & \frac{0}{7} \\ \frac{0}{7} & \frac{7}{7} \end{bmatrix} \begin{bmatrix} x \\ y \end{bmatrix} = \begin{bmatrix} \frac{14}{7} \\ \frac{^-7}{7} \end{bmatrix}$$

$$\begin{bmatrix} 1 & 0 \\ 0 & 1 \end{bmatrix} \begin{bmatrix} x \\ y \end{bmatrix} = \begin{bmatrix} 2 \\ -1 \end{bmatrix}$$

$$1*x + 0*y = 2$$
$$x = 2$$
$$0*x + 1*y = ^-1$$
$$y = ^-1$$

(b)   Solve for x and y

$$x + y = 3$$
$$3x - 2y = ^-1$$

Solution

$$x + y = 3$$
$$3x - 2y = ^-1$$

In matrix it becomes

$$\begin{bmatrix} 1 & 1 \\ 3 & -2 \end{bmatrix} \begin{bmatrix} x \\ y \end{bmatrix} = \begin{bmatrix} 3 \\ -1 \end{bmatrix} \qquad (i)$$

The inverse $= \dfrac{1}{(1*^-2)-(3*1)} \begin{bmatrix} ^-2 & ^-1 \\ ^-3 & 1 \end{bmatrix}$

$$= \frac{1}{^-2 - 3}\begin{bmatrix} ^-2 & ^-1 \\ ^-3 & 1 \end{bmatrix}$$

$$= \frac{1}{^-5}\begin{bmatrix} ^-2 & ^-1 \\ ^-3 & 1 \end{bmatrix}$$

$$= -\frac{1}{5}\begin{bmatrix} ^-2 & ^-1 \\ ^-3 & 1 \end{bmatrix}$$

$$= \begin{bmatrix} \dfrac{2}{5} & \dfrac{1}{5} \\ \dfrac{3}{5} & \dfrac{^-1}{5} \end{bmatrix}$$

Multiply the inverse on both sides of equation *(i)*

$$\begin{bmatrix} \dfrac{2}{5} & \dfrac{1}{5} \\ \dfrac{3}{5} & \dfrac{^-1}{5} \end{bmatrix}\begin{bmatrix} 1 & 1 \\ 3 & ^-2 \end{bmatrix}\begin{bmatrix} x \\ y \end{bmatrix} = \begin{bmatrix} \dfrac{2}{5} & \dfrac{1}{5} \\ \dfrac{3}{5} & \dfrac{^-1}{5} \end{bmatrix}\begin{bmatrix} 3 \\ ^-1 \end{bmatrix}$$

$$\begin{bmatrix} \dfrac{2}{5}*1 + \dfrac{1}{5}*3 & \dfrac{2}{5}*1 + \dfrac{1}{5}*{}^-2 \\ \dfrac{3}{5}*1 + \dfrac{^-1}{5}*3 & \dfrac{3}{5}*1 + \dfrac{^-1}{5}*{}^-2 \end{bmatrix}\begin{bmatrix} x \\ y \end{bmatrix} = \begin{bmatrix} \dfrac{2}{5}*3 + \dfrac{1}{5}*{}^-1 \\ \dfrac{3}{5}*3 + \dfrac{^-1}{5}*{}^-1 \end{bmatrix}$$

$$\begin{bmatrix} \dfrac{2}{5} + \dfrac{3}{5} & \dfrac{2}{5} - \dfrac{2}{5} \\ \dfrac{3}{5} - \dfrac{3}{5} & \dfrac{3}{5} + \dfrac{2}{5} \end{bmatrix}\begin{bmatrix} x \\ y \end{bmatrix} = \begin{bmatrix} \dfrac{6}{5} - \dfrac{1}{5} \\ \dfrac{9}{5} + \dfrac{1}{5} \end{bmatrix}$$

$$\begin{bmatrix} \dfrac{5}{5} & \dfrac{0}{5} \\ \dfrac{0}{5} & \dfrac{5}{5} \end{bmatrix}\begin{bmatrix} x \\ y \end{bmatrix} = \begin{bmatrix} \dfrac{5}{5} \\ \dfrac{10}{5} \end{bmatrix}$$

$$\begin{bmatrix} 1 & 0 \\ 0 & 1 \end{bmatrix}\begin{bmatrix} x \\ y \end{bmatrix} = \begin{bmatrix} 1 \\ 2 \end{bmatrix}$$

$$1*x + 0*y = 1$$
$$x = 1$$
$$0*x + 1*y = 2$$
$$y = 2$$

Example 11.4

In a food store there is a fruit basket containing 18 fruits, mangoes and oranges, each costing $5 and $3 respectively. The total cost of all fruits in the basket is $70. Find the total number of mangoes in the basket.

Solution

Let the total number of mangoes in the basket be m
Let the total number of oranges in the basket be n
The total cost for all mangoes $\rightarrow 5 \times m = 5m$
The total cost for all oranges $\rightarrow 3 \times n = 3n$

$$m + n = 18 \qquad \text{(i)}$$
$$5m + 3n = 70 \qquad \text{(ii)}$$

Multiply by 3 in equation (i)
$$3(m + n = 18)$$
$$3m + 3n = 54 \qquad \text{(iii)}$$

Subtract equation (ii) − (iii)

$$
\begin{array}{rl}
5m + 3n = 70 & (ii) \\
- \quad 3m + 3n = 54 & (iii) \\
\hline
2m = 16 &
\end{array}
$$

$$\frac{\cancel{2}m}{\cancel{2}} = \frac{\cancel{16}^{8}}{\cancel{2}}$$
$$m = 8$$

∴ the total number of mangoes in the basket is 8

Exercise 11A

Use one of the three methods to solve for x and y
for numbers 1 to 7.

(1) $x + \frac{1}{2}y = 2$

        $2x - 4y = {}^-11$

(2) $x + y = 3$

        $7x - 2y = {}^-24$

(3) $7x + 2y = 15$

        $2x - 6y = 24$

(4) $11x + y = {}^-1$

        $5x + 12y = {}^-12$

(5) $2x + 9y = 31$

        $3x + 2y = 12$

(6) $2x + y = 46$

        $x - \frac{1}{2}y = 3$

(7) $5x + 10y = 2$

        $-\frac{1}{3}x + \frac{8}{5}y = 1$

(8) In a book store there is a box containing 35 text books of
Math and English, each costing $20 and $10 respectively.
The total cost of all the text books in the box is $550.
Find the total number of Math text books in the box.

(9) Festo has a building of 40 apartments; 1 bed room
apartments, 2 bed room apartments and 3 bed room
apartments and their monthly payments are $900, $1000
and $1400 respectively. The total monthly payment of
all the apartments in the building is $40000. If there is
no payment from all the 3 bed room apartments, the total
monthly payment of the other apartments in the building
comes down to $33000. Find the total number of 2 bed
room apartments in the building.

SOLUTIONS FOR EXERCISE 11A

(1)
$$x + \frac{1}{2}y = 2$$
$$2x - 4y = {}^-11$$

$$x + \frac{1}{2}y = 2 \qquad \text{(i)}$$
$$2x - 4y = {}^-11 \qquad \text{(ii)}$$

Multiply equation *(i)* by 2

$$2(x + \frac{1}{2}y = 2)$$
$$2x + y = 4 \qquad \text{(iii)}$$

Subtract equation *(iii)* from *(ii)*

$$
\begin{array}{ll}
2x - 4y = {}^-11 & (ii) \\
- \quad 2x + y = 4 & (iii) \\
\hline
{}^-5y = {}^-15 &
\end{array}
$$

Divide both sides by ${}^-5$

$$\frac{{}^-5y}{{}^-5} = \frac{{}^-15^3}{{}^-5}$$

$$y = 3$$

From equation *(i)*

$$x + \frac{1}{2}y = 2$$

$$x + \frac{1}{2} * 3 = 2$$

$$x + \frac{3}{2} = 2$$

$$x = 2 - \frac{3}{2}$$

$$x = \frac{4 - 3}{2}$$

$$x = \frac{1}{2}$$

(2) $x + y = 3$
$7x - 2y = {}^-24$

$$x + y = 3 \qquad \qquad \text{(i)}$$
$$7x - 2y = {}^-24 \qquad \qquad \text{(ii)}$$

Multiply equation *(i)* by 2
$$2(x + y = 3)$$
$$2x + 2y = 6 \qquad \qquad \text{(iii)}$$

Add equation *(ii)* and *(iii)*

$$7x - 2y \ = \ {}^-24 \qquad \qquad (ii)$$
$$+ \ 2x + 2y \ = \ 6 \qquad \qquad (iii)$$
$$\overline{}$$
$$9x \ \quad = \ {}^-18$$

Divide both sides by 9
$$\frac{\cancel{9}x}{\cancel{9}} = \frac{{}^-\cancel{18}^2}{\cancel{9}}$$
$$x = {}^-2$$

From equation *(i)*
$$x + y = 3$$
$${}^-2 + y = 3$$
$$y = 3 + 2$$
$$y = 5$$

(3) $7x + 2y = 15$
$2x - 6y = 24$

$$7x + 2y = 15 \qquad (i)$$
$$2x - 6y = 24 \qquad (ii)$$

Multiply equation *(i)* by 3
$$3(7x + 2y = 15)$$
$$21x + 6y = 45 \qquad (iii)$$

Add equation *(ii)* and *(iii)*

$$2x - 6y = 24 \qquad (ii)$$
$$+\ 21x + 6y = 45 \qquad (iii)$$
$$\overline{\qquad\qquad\qquad\qquad}$$
$$23x \qquad = 69$$

Divide both sides by 23
$$\frac{\cancel{23}x}{\cancel{23}} = \frac{\cancel{69}^3}{\cancel{23}}$$
$$x = 3$$

From equation (i)
$$7x + 2y = 15$$
$$7*3 + 2y = 15$$
$$21 + 2y = 15$$
$$2y = 15 - 21$$
$$2y = {}^-6$$
$$\frac{\cancel{2}y}{\cancel{2}} = \frac{\cancel{{}^-6}^{\ {}^-3}}{\cancel{2}}$$
$$y = {}^-3$$

(4) $\qquad 11x + y = {}^-1$
$\qquad\quad\ \ 5x + 12y = {}^-12$

$$11x + y = {}^-1 \qquad (i)$$
$$5x + 12y = {}^-12 \qquad (ii)$$

Multiply equation *(i)* by 12
$$12(11x + y = {}^-1)$$
$$132x + 12y = {}^-12 \qquad (iii)$$

Subtract equation *(ii) − (iii)*

$$5x + 12y = {}^-12 \qquad (ii)$$
$$+\ 132x + 12y = {}^-12 \qquad (iii)$$

$$\overline{{}^-127x \qquad\ = 0}$$

Divide both sides by *127*

$$\frac{\cancel{127}x}{\cancel{127}} = \frac{0}{\cancel{127}}$$
$$x = 0$$

From equation (i)

$$11x + y = {}^-1$$
$$11*0 + y = {}^-1$$
$$0 + y = {}^-1$$
$$y = {}^-1$$

(5) $\qquad 2x + 9y = 31$

$\qquad 3x + 2y = 12$

$$2x + 9y = 31 \qquad (i)$$
$$3x + 2y = 12 \qquad (ii)$$

Multiply equation *(ii)* by $\frac{2}{3}$

$$\frac{2}{3}(3x + 2y = 12)$$

$$2x + \frac{4}{3}y = 8 \qquad (iii)$$

Subtract equation *(i) − (iii)*

$$2x + 9y = 31 \qquad (i)$$
$$+\ 2x + \frac{4}{3}y = 8 \qquad (iii)$$

$$\overline{\frac{23}{3}y\ = 23}$$

Multiply both sides by $\dfrac{3}{23}$

$$\dfrac{3}{23} \times \dfrac{23}{3} y = 23 \times \dfrac{3}{23}$$
$$y = 3$$

From equation (ii)
$$3x + 2y = 12$$
$$3x + 2*3 = 12$$
$$3x + 6 = 12$$
$$3x = 12 - 6$$
$$3x = 6$$
$$\dfrac{\cancel{3}x}{\cancel{3}} = \dfrac{\cancel{6}^2}{\cancel{3}}$$
$$x = 2$$

(6) $$2x + y = 46$$
$$x - \dfrac{1}{2}y = 3$$

$$2x + y = 46 \qquad\qquad (i)$$
$$x - \dfrac{1}{2}y = 3 \qquad\qquad (ii)$$

Multiply equation (ii) by 2
$$2(x - \dfrac{1}{2}y = 3)$$
$$2x - y = 6 \qquad\qquad (iii)$$

Add equation (i) and (iii)
$$\begin{array}{rl} 2x + y = 46 & (i) \\ + \; 2x - y = 6 & (iii) \\ \hline 4x \qquad = 52 & \end{array}$$

Divide both sides by 4
$$\dfrac{4x}{4} = \dfrac{\cancel{52}^{13}}{4}$$

224

$$x = 13$$

From equation (i)

$$2x + y = 46$$
$$2*13 + y = 46$$
$$26 + y = 46$$
$$y = 46 - 26$$
$$y = 20$$

(7) $$5x + 10y = 2$$
$$-\frac{1}{3}x + \frac{8}{5}y = 1$$

$$5x + 10y = 2 \qquad\qquad\qquad (i)$$
$$-\frac{1}{3}x + \frac{8}{5}y = 1 \qquad\qquad\qquad (ii)$$

Multiply equation (i) by $\dfrac{1}{15}$

$$\frac{1}{15}(5x + 10y = 2)$$
$$\frac{1}{3}x + \frac{2}{3}y = \frac{2}{15} \qquad\qquad (iii)$$

Add equation (ii) and (iii)

$$-\frac{1}{3}x + \frac{8}{5}y = 1 \qquad\qquad (ii)$$
$$+\ \frac{1}{3}x + \frac{2}{3}y = \frac{2}{15} \qquad\qquad (iii)$$
$$\overline{\qquad\qquad\qquad\qquad\qquad}$$
$$\frac{34}{15}y = \frac{17}{15}$$

Multiply both sides by $\dfrac{15}{34}$

$$\frac{15}{34} \times \frac{34}{15}y = \frac{17}{15} \times \frac{15}{34}$$
$$y = \frac{1}{2}$$

From equation *(i)*

$$5x + 10y = 2$$

$$5x + 10 * \frac{1}{2} = 2$$

$$5x + 5 = 2$$

$$5x = 2 - 5$$

$$5x = {}^-3$$

$$\frac{5x}{5} = \frac{{}^-3}{5}$$

$$x = -\frac{3}{5}$$

(8) Let the total number of Math text books in a box be x

Let the total number of English text books in a box be y

The total cost for all Math text books → *20 * x = 20x*

The total cost for all English text books → *10 * y = 10y*

$$x + y = 35 \qquad \text{(i)}$$

$$20x + 10y = 550 \qquad \text{(ii)}$$

Multiply by *10* in equation *(i)*

$$10(x + y = 35)$$

$$10x + 10y = 350 \qquad \text{(iii)}$$

Subtract equation (ii) − (iii)

$$20x + 10y = 550 \qquad (ii)$$

$$- \quad 10x + 10y = 350 \qquad (iii)$$

$$\overline{}$$

$$10x \qquad = 200$$

$$\frac{10x}{10} = \frac{200^{20}}{10}$$

$$x = 20$$

∴ the total number of Math text books in a box is *20*.

(9) Let the total number of *1* bed room apartments be x

Let the total number of *2* bed room apartments be y

Let the total number of 3 bed room apartment be z

$$x + y + z = 40 \qquad \text{(i)}$$
$$900x + 1000y + 1400z = 40000 \qquad \text{(ii)}$$
$$900x + 1000y = 33000 \qquad \text{(iii)}$$

Subtract equation *(ii) – (iii)*

$$1400z = 40000 - 33000$$
$$1400z = 7000$$
$$\frac{\cancel{1400}z}{\cancel{1400}} = \frac{\cancel{7000}^{5}}{\cancel{1400}}$$
$$z = 5$$

Multiply *1000* in equation *(i)*

$$1000(x + y + z = 40)$$
$$1000x + 1000y + 1000z = 40000 \qquad \text{(iv)}$$

Subtract equation *(iv) – (ii)*

$$1000x + 1000y + 1000z = 40000 \qquad (iv)$$
$$- \quad 900x + 1000y + 1400z = 40000 \qquad (ii)$$

$$\overline{\qquad\qquad\qquad\qquad\qquad\qquad\qquad}$$

$$100x - 400z = 0 \qquad (v)$$

Substitute $z = 5$ in equation *(v)*

$$100x - 400 * 5 = 0$$
$$100x - 2000 = 0$$
$$100x = 2000$$
$$\frac{\cancel{100}x}{\cancel{100}} = \frac{\cancel{2000}^{20}}{\cancel{100}}$$
$$x = 20$$

Substitute $x = 20$ in equation *(iii)*

$$900 * 20 + 1000y = 33000$$
$$18000 + 1000y = 33000$$
$$1000y = 33000 - 18000$$
$$1000y = 15000$$
$$\frac{\cancel{1000}y}{\cancel{1000}} = \frac{\cancel{15000}^{15}}{\cancel{1000}}$$
$$y = 15$$

∴ the total number of 2 bed room apartments is *15*

Exercise 11B

Use one of the three methods to solve for x and y
for numbers 1 to 9.

(1) $2x + \dfrac{1}{2}y = 3$

   $x - 7y = {}^-13$

(2) $x + 2y = 10$

   $3x - 4y = {}^-30$

(3) ${}^-5x + 9y = 14$

   $2x - 6y = {}^-8$

(4) $9x + 2y = 7$

   $3x + 7y = 15$

(5) $2x + 9y = {}^-1$

   $8x + 12y = {}^-4$

(6) $2x - y = 11$

   $3x + 3y = 30$

(7) $4x + 3y = 5$

   $3x + 4y = {}^-5$

(8) $x + 2y = 35$

   $2x - \dfrac{1}{2}y = 16$

(9) $6x + 4y = 10$

   ${}^-\dfrac{1}{5}x + \dfrac{8}{10}y = {}^-5$

(10) In book store there is a box containing 24 text books of Math and English , each costing $10 and $5 respectively. The total cost of all the text books in the box is $200. Find the total number of Math text books in the box.

(11) David has a building of 20 apartments; 1 bed room apartments, 2 bed room apartments and 3 bed room apartments and their monthly payments are $500, $800 and $1000 respectively. The total monthly payment of all the apartments in the building is $15500. If there is

no payment from all the 3 bed room apartments, the total monthly payment of the other apartments in the building comes down to $10500. Find the total number of 2 bed room apartments in the building.

CHAPTER 12

CO-ORDINATE, FUNCTION, EQUATION AND GRAPH

12-1 Coordinates, a Relation and a Function

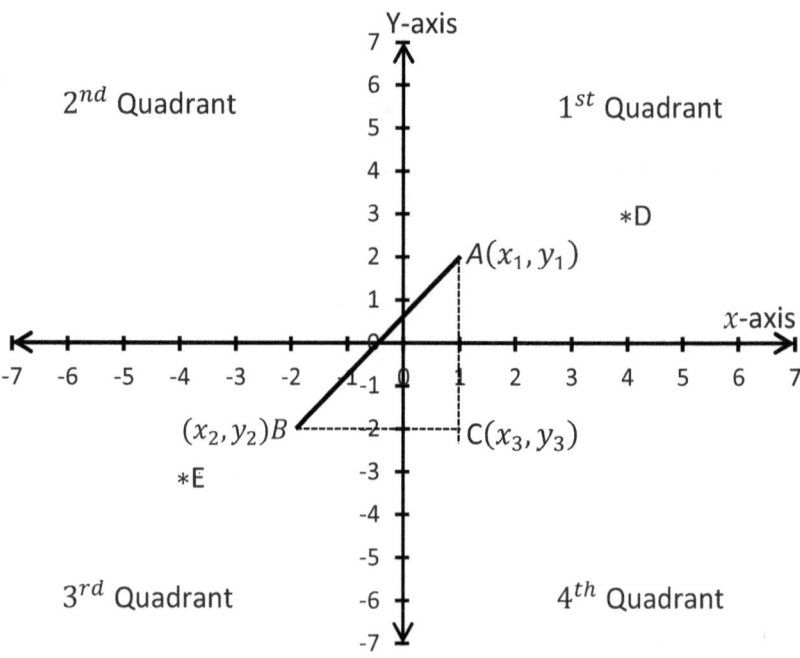

Figure 12.1

The graph in figure *12.1* above shows <u>Cartesian Coordinate Plane</u> which has horizontal axis called <u>x-axis</u> (*y=0*), and the vertical axis called <u>y-axis</u> (*x=0*), the intersection of these two is <u>origin</u> (0,0)

Any point on a graph is identified by ordered pair (*x, y*),

x-coordinate written first and then y- coordinate. Points D and E on the graph are written as D (4, 3) found in 1st quadrant and E (-4, -3) found in 3rd quadrant.

Generally the value of the variable y depends on the value of the variable x, therefore y is the dependent variable and x is the independent variable.

$$(x, y)$$

independent variable ↗   ↖ *dependent variable*

Relation is a set of ordered pairs such as *(2, 4), (4, 8), (8, 16)*

A Function is a relation which has exactly one value of the dependent variable for each value of the independent variable.

Examples 12.1

(a)   Given A = (3, -1), (5, -3), (7, -5)

B = (1, 4), (0, 3), (0, 4)

C = (1, 6), (6, 6)

A and C are functions, for each x-value, there is exactly one y-value in other words all x-values of A and all the x-values of C are different from each other.

Though C has same y-value (6) it is still a Function.

B is not a Function, the last two ordered pair has same x-value (0)

So B is a relation but not a Function

Domain is the set of all values of independent variable (x)

Range is the set of all values of the dependent variable (y)

Given the Domain and the Range in **Examples 12.1(a)** above, we can define a function.

From A = (3, -1), (5, -3), (7, -5)

Domain = { 3, 5, 7 }

The Range = { -1, -3, -5 }

A is a Function because domain has all its values different from each other.

    From B = (1, 4), (0, 3), (0, 4)
    Domain = { 1, 0, 0 }
    The Range = { 4, 3, 4 }

B is not a Function because Domain has two same values of (0)

A Relation, a Function, a Domain and a Range are ordered pairs which can be represented on tables and graphs as shown below.

For C = (1, 6), (6, 6)

x	y
1	6
6	6

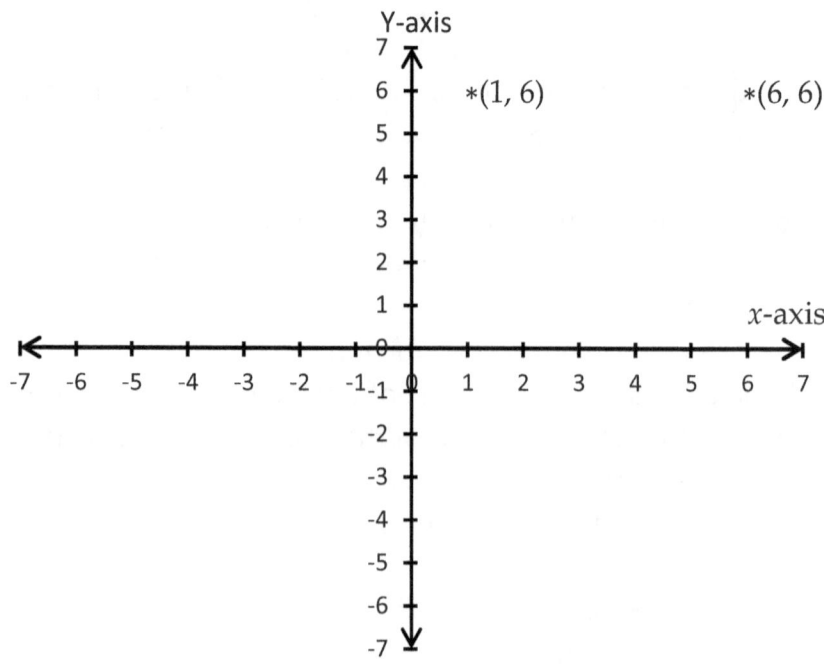

Examples 12.1

Given

$A = (1, {}^-3), (3, {}^-5), (5, {}^-7)$

$B = (3, 6), (2, 5), (2, 4)$

$C = ({}^-1, 3), ({}^-1, 6), (0, 4)$

$D = (6, 0), (0, 6)$

b(i) Find the Domain and the Range of A, B, C and D

For A Domain$(x) = \{ 1, 3, 5 \}$
          Range $(y) = \{ {}^-3, {}^-5, {}^-7 \}$

For B Domain$(x) = \{ 3, 2, 2 \}$
          Range $(y) = \{ 6, 5, 4 \}$

For C Domain$(x) = \{ {}^-1, {}^-1, 0 \}$
          Range $(y) = \{ 3, 6, 4 \}$

For D Domain$(x) = \{ 6, 0 \}$
          Range $(y) = \{ 0, 6 \}$

(ii) Find weather A, B, C and D are Function.

A is a Function because all its Domain or x-values are different from each other.

B is not a Function because it has the same Domain or x-values which are 2.

C is not a Function because it has the same Domain or x-values which are $^-1$.

D is a Function because all its Domain or x-values are different from each other.

(iii) Make a table and a graph for D

For $D = (6, 0), (0, 6)$

x	y
6	0
0	6

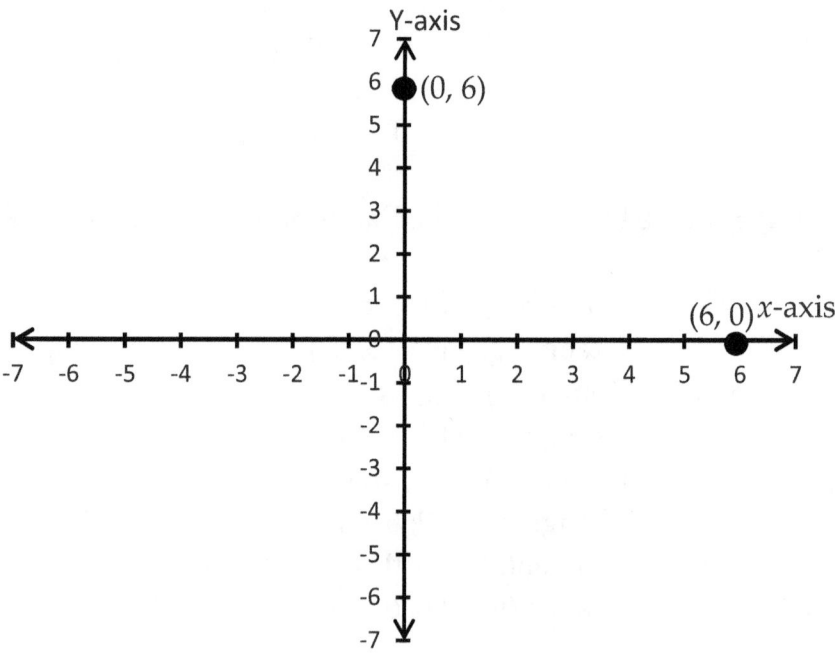

12-2 Operations on Function Notation

Given the Function $f_{(x)} = ax + b$

$$g_{(x)} = x^2$$

To find $h_{(x)} = g \circ f_{(x)}$ read as g composed with f or read as g of $f_{(x)}$

$$h_{(x)} = g(f_{(x)}) = (f_{(x)})^2 = (ax + b)^2 \qquad \text{OR}$$
$$h_{(x)} = g(f_{(x)}) = g(ax + b) = (ax + b)^2$$

Note that $f \circ g_{(x)} = f(g_{(x)}) = f(x^2) = ax^2 + b$

Therefore $g \circ f_{(x)} \neq f \circ g_{(x)}$

Examples 12.2

(a) Given $f_{(x)} = 3x + 7$

(i) Find $f_{(3)}$

Solution

$$f_{(3)} = 3(3) + 7$$
$$= 9 + 7$$
$$f_{(3)} = 16$$

(ii) Given $g_{(x)} = 2x + 1$

Find $g_{(a+1)}$

Solution

$$g_{(a+1)} = 2(a + 1) + 1$$
$$= 2a + 2 + 1$$
$$g_{(a+1)} = 2a + 3$$

(b) Use Function notation equation to find $f_{(1)}$ and $f_{(a)}$

$$x^2 - y = {}^-1$$

Solution

Make y the subject by subtracting x^2 on both sides

$$x^2 - x^2 - y = {}^-1 - x^2$$
$${}^-y = {}^-1 - x^2$$

Multiply or divide by -1 on both sides

$$y = 1 + x^2$$
$$f_{(x)} = 1 + x^2$$
$$f_{(x)} = x^2 + 1$$
$$f_{(1)} = 1^2 + 1$$
$$f_{(1)} = 2$$

From $f_{(x)} = x^2 + 1$

$$f_{(a)} = a^2 + 1$$

(c) Given the Function

$$f_{(x)} = 3x + 2 \quad \text{and}$$
$$g_{(x)} = x^2$$

Find:

(i) $\dfrac{f(x+h) - f(x)}{h}$

Solution

$$\text{From } f_{(x)} = 3x + 2$$
$$f_{(x+h)} = 3(x+h) + 2$$
$$= 3x + 3h + 2$$
$$\therefore \frac{f(x+h) - f(x)}{h} = \frac{3x + 3h + 2 - (3x+2)}{h}$$

$$= \frac{3x + 3h + 2 - 3x - 2}{h}$$

$$= \frac{3\cancel{h}}{\cancel{h}}$$
$$= 3$$

(ii) $g_{(x)} + f_{(x)}$

Solution

$$g_{(x)} + f_{(x)} = x^2 + 3x + 2$$
$$\text{Multiply } x^2 \text{ and } 2$$

$$2x^2 \diagup\!\!\!\!\diagdown \begin{array}{c} x \\ 2x \end{array}$$

$$\text{a pair or 2 factors whose sum is } 3x$$
$$x^2 + x + 2x + 2$$
$$x(x+1) + 2(x+1)$$
$$g_{(x)} + f_{(x)} = (x+1)(x+2)$$

(d) Find the following inverse functions given $f_{(x)} = x + 5$

(i) $f_{(x)}^{-1}$

Solution

$$f_{(x)} = x + 5$$
$$y = x + 5$$
$$\text{Interchange } x \text{ and } y$$
$$x = y + 5$$

Make y the subject by subtracting 5 on both sides

$$y = x - 5$$
$$\therefore f^{-1}_{(x)} = x - 5$$

(ii) $f \, o \, f^{-1}_{(x)}$

Solution

$$f \, o \, f^{-1}_{(x)} = f_{(x-5)}$$
$$= x - 5 + 5$$
$$= x$$

(iii) $f^{-1} \, o \, f_{(x)}$

Solution

$$f^{-1} \, o \, f_{(x)} = f^{-1}_{(x+5)}$$
$$= x + 5 - 5$$
$$= x$$
$$\therefore \quad f \, o \, f^{-1}_{(x)} = f^{-1} \, o \, f_{(x)} = x$$

12-3 Coordinates, equations and graphs

Examples 12.3

a(i) Find a distance between A and B in the graph in figure *12.1*.

Solution

The distance (d) between two points is given by

$$d = \sqrt{(x_1 - x_2)^2 + (y_1 - y_2)^2}$$

From the graph in figure *12.1* distance

$$AB = \sqrt{(x_1 - x_2)^2 + (y_1 - y_2)^2}$$
$$AB = \sqrt{(1 - {}^-2)^2 + (2 - {}^-2)^2}$$
$$AB = \sqrt{(1 + 2)^2 + (2 + 2)^2}$$
$$AB = \sqrt{3^2 + 4^2}$$

$$AB = \sqrt{9 + 16}$$
$$AB = \sqrt{25}$$
$$= \sqrt{5^2}$$
$$AB = 5$$

$\therefore$ The distance AB is 5

The Pythagorean formula

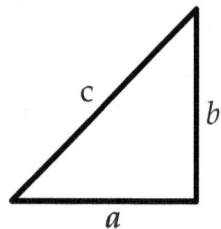

From a right angle triangle

$$a^2 + b^2 = c^2$$

We can also get distance AB from the graph in figure 12.1 using a Pythagorean formula

From $c^2 = a^2 + b^2$

$A\,(x_1 y_1)$

$$AB^2 = BC^2 + AC^2$$

$(x_2 y_2)B$ $C\,(x_3 y_3)$

$$AB^2 = BC^2 + AC^2$$
$$= (x_3 - x_2)^2 + (y_3 - y_2)^2 + (x_1 - x_3)^2 + (y_1 - y_3)^2$$
$$= (1 - {}^-2)^2 + ({}^-2 - {}^-2)^2 + (1 - 1)^2 + (2 - {}^-2)^2$$
$$= (1 + 2)^2 + ({}^-2 + 2)^2 + (1 - 1)^2 + (2 + 2)^2$$
$$= 3^2 + 0^2 + 0^2 + 4^2$$
$$= 9 + 16$$
$$AB^2 = 25$$

Introduce square roots on both sides
$$\sqrt{AB^2} = \sqrt{25}$$
$$\sqrt{AB^2} = \sqrt{5^2}$$

$$AB = 5$$
∴ the distance AB is 5

(ii) Find a midpoint of a distance or a line $\overline{AB}$ in figure *12.1*
Solution

Midpoint formula is $\left(\dfrac{x_1 + x_2}{2} , \dfrac{y_1 + y_2}{2} \right)$

Midpoint of $\overline{AB} = \left(\dfrac{1 + {}^-2}{2} , \dfrac{2 + {}^-2}{2} \right)$

$= \left(\dfrac{1 - 2}{2} , \dfrac{2 - 2}{2} \right)$

$= \left(\dfrac{{}^-1}{2} , \dfrac{0}{2} \right)$

Midpoint of $\overline{AB} = \left(\dfrac{{}^-1}{2} , 0 \right)$

(iii) If the Domain of the points on a line $\overline{AB}$ is { 2, 4, 5, 9 },
find their Mean.
Solution

The Mean $M = \dfrac{x_1 + x_2 + \ldots\ldots\ldots + x_n}{n}$

$= \dfrac{2 + 4 + 5 + 9}{4}$

$= \dfrac{20}{4}$

$= \dfrac{\overset{5}{\cancel{20}}}{4}$

$= 5$

∴ their Mean is 5

b(i) Find the slope of the line $\overline{AB}$ from figure 12.1

Solution

$$\text{Slope formula } m = \frac{rise}{run} = \frac{\Delta y}{\Delta x} = \frac{y_2 - y_1}{x_2 - x_1} \quad \text{where } x_2 \neq x_1$$

$$\text{Slope of } \overline{AB} = \frac{{}^-2 - 2}{{}^-2 - 1}$$

$$= \frac{{}^-4}{{}^-3} = \frac{4}{3}$$

$$\text{Slope of } \overline{AB} = \frac{4}{3}$$

(ii) Given from figure 12.1 $(x_1, y_1) = (1, 2)$ and $m = \frac{4}{3}$.

Find the equation of the line $\overline{AB}$ and the y-intercept.

Solution

From the formula of the slope, we get the equation of the line $\quad$ **y = mx + c** $\quad$ **where c is y-intercept (0, c)**

Given the slope m and the point on the line $(x_1 y_1)$

$$m = \frac{y - y_1}{x - x_1} \quad \text{or} \quad m(x - x_1) = y - y_1$$

$$\frac{4}{3} = \frac{y - 2}{x - 1}$$

Cross multiply

$$3(y - 2) = 4(x - 1)$$

Divide both sides by 3

$$\frac{3(y - 2)}{3} = \frac{4(x - 1)}{3}$$

$$y - 2 = \frac{4}{3}(x - 1)$$

$$y - 2 = \frac{4x}{3} - \frac{4}{3}$$

add 2 to both sides of equation

$$y - 2 + 2 = \frac{4x}{3} - \frac{4}{3} + 2$$

$$y = \frac{4x}{3} + \frac{^-4 + 6}{3}$$

$$y = \frac{4x}{3} + \frac{2}{3}$$

∴ The equation of the line $\overline{AB}$ is $y = \frac{4x}{3} + \frac{2}{3}$

Compare with equation of the line formula $y = mx + c$

$$m = \frac{4}{3} \quad \text{and} \quad c = \frac{2}{3} \quad \text{y-intercept} \left(0, \frac{2}{3} \right)$$

(iii) Find two other points on the line $\overline{AB}$.

Solution

Using the equation of the line $\overline{AB}$

$$y = \frac{4x}{3} + \frac{2}{3}$$

When $x = 0$

$$y = \frac{4 \times 0}{3} + \frac{2}{3}$$

$$y = 0 + \frac{2}{3}$$

$$y = \frac{2}{3}$$

When $x = 0$, $y = \frac{2}{3}$.

Therefore one of the points is $\left(0, \frac{2}{3} \right)$

when $y = 0$

$$0 = \frac{4x}{3} + \frac{2}{3}$$

$$\frac{4x}{3} = -\frac{2}{3}$$

Multiply both sides by $\frac{3}{4}$

$$\frac{3}{4} \times \frac{4x}{3} = -\frac{2}{3} \times \frac{3}{4}$$

$$x = -\frac{1}{2}$$

When $y = 0$, $x = -\frac{1}{2}$.

Therefore the second point is $\left(-\frac{1}{2}, 0\right)$

(iv) Find the slope of the line perpendicular to $\overline{AB}$ in the graph in figure 12.1

Solution

The product of the slope of perpendicular lines is -1.
Therefore $m_1 m_2 = -1$

Slope of $\overline{AB} = m_1 = \frac{4}{3}$

$$m_1 m_2 = -1$$
$$\frac{4}{3} m_2 = -1$$

Multiply both sides by $\frac{3}{4}$

$$\frac{3}{4} \times \frac{4}{3} m_2 = {}^-1 \times \frac{3}{4}$$

$$m_2 = -\frac{3}{4}$$

Therefore the slope of a line perpendicular to $\overline{AB}$ is $-\dfrac{3}{4}$

<u>Note</u>

the slope of parallel lines are equal, $m = m_1 = m_2$
the slope of horizontal line is $0, m = 0, for\ example\ line\ y = 0$
the slope of vertical line is undefined, $m = undefined, x = 0$

Examples 12.4

Given two points $({}^-2, 0)$ and $(1, 4)$. Find the following:

(i) The distance between the two points

Solution

$$\begin{aligned}
\text{From} \quad d &= \sqrt{(x_1 - x_2)^2 + (y_1 - y_2)^2} \\
&= \sqrt{({}^-2 - 1)^2 + (0 - 4)^2} \\
&= \sqrt{{}^-3^2 + {}^-4^2} \\
&= \sqrt{9 + 16} \\
&= \sqrt{25} \\
&= 5
\end{aligned}$$

∴ the distance between the two points is 5

(ii) The midpoint of the line between the two points

Solution

$$\text{Midpoint formula} = \left(\frac{x_1 + x_2}{2}, \frac{y_1 + y_2}{2}\right)$$

$$= \left(\frac{{}^-2 + 1}{2}, \frac{0 + 4}{2}\right)$$

$$= \left(\frac{{}^-1}{2}, \frac{4}{2}\right)$$

$$\text{Midpoint} = \left(-\frac{1}{2}, 2\right)$$

(iii) Slope of the line between the two points
Solution

$$\text{From slope } m = \frac{y_2 - y_1}{x_2 - x_1}$$

$$= \frac{4 - 0}{1 - {}^-2}$$

$$= \frac{4}{1 + 2}$$

$$= \frac{4}{3}$$

(iv) The equation of the line between the two points
Solution

$$\text{From } m = \frac{y - y_1}{x - x_1}$$

$$\frac{4}{3} = \frac{y - 0}{x - {}^-2}$$

$$\frac{4}{3} = \frac{y}{x + 2}$$

Cross multiply

$$3y = 4(x + 2)$$

Divide both sides by 3

$$\frac{3y}{3} = \frac{4}{3}(x + 2)$$

$$y = \frac{4}{3}(x + 2)$$

$$y = \frac{4}{3}x + \frac{8}{3}$$

That's the equation of the line between the two points

(v) The slope of the line parallel to a line between the two points

For parallel lines $m = m_1 = \frac{4}{3}$

(vi) The slope of the line perpendicular to a line between the two points

Solution

For perpendicular lines

$$m_1 m_2 = {}^{-}1$$

$$\frac{4}{3} m_2 = {}^{-}1$$

Multiply both sides by 3

$$3 \times \frac{4}{3} m_2 = {}^{-}1 \times 3$$

$$4m_2 = {}^{-}3$$

Divide both sides by 4

$$\frac{4m_2}{4} = \frac{{}^{-}3}{4}$$

$$m_2 = \frac{{}^{-}3}{4}$$

∴ the slope of the line perpendicular to a line between the two points is $\frac{{}^{-}3}{4}$

(vii) The x-intercept and y-intercept

Solution

x-intercept is the point on the line where $y = 0$

From the equation of the line

$$y = \frac{4}{3}x + \frac{8}{3}$$

$$0 = \frac{4}{3}x + \frac{8}{3}$$

Subtract $\frac{8}{3}$ from both sides

$$0 - \frac{8}{3} = \frac{4}{3}x + \frac{8}{3} - \frac{8}{3}$$

$$\frac{^-8}{3} = \frac{4}{3}x$$

Multiply both sides by $\frac{3}{4}$

$$\frac{^-8}{3} \times \frac{3}{4} = \frac{3}{4} \times \frac{4}{3}x$$

$$^-2 = x$$

$\therefore$ x-intercept is ($^-2, 0$)

y-intercept is a point on the line where $x = 0$

Also from the equation of the line

$$y = \frac{4}{3}x + \frac{8}{3}$$

$$y = \frac{4}{3} \times 0 + \frac{8}{3}$$

$$y = 0 + \frac{8}{3}$$

$$y = \frac{8}{3}$$

$$\therefore y\text{-intercept is } \left(0, \frac{8}{3}\right)$$

Or compare the equation of the line $y = \frac{4}{3}x + \frac{8}{3}$ with

$$y = mx + c$$

$$c = \frac{8}{3} \text{ at } y\text{-intercept } \left(0, \frac{8}{3}\right)$$

There are two ways of writing the equation of the line:
A slope intercept form $y = mx + c$, and a standard form
$ax + by = c$

Examples 12.5

(a) Find the equation of a line in a standard form $ax + by = c$

Given the slope is $\frac{^{-}1}{3}$ and y-intercept is 5.

Solution

From $y = mx + c$

$$y = \frac{^{-}1}{3}x + 5$$

Add $\frac{1}{3}x$ to both sides of equation

$$\frac{1}{3}x + y = \frac{^{-}1}{3}x + \frac{1}{3}x + 5$$

$$\frac{1}{3}x + y = 5$$

Multiply both sides by 3

$$3\left(\frac{1}{3}x + y\right) = 5 \times 3$$

$$x + 3y = 15$$

(b) Find the equation of a line in a slope intercept form

$y = mx + c$

given the slope is 3 and a point on a line is (3, 8)

Solution

From slope $m = \dfrac{y - y_1}{x - x_1}$

$3 = \dfrac{y - 8}{x - 3}$

Cross multiply

$y - 8 = 3(x - 3)$

$y - 8 = 3x - 9$

Add 8 to both sides

$y - 8 + 8 = 3x - 9 + 8$

$y = 3x - 1$

Examples 12.6

Find r given two points and slope of the line:

(i)   (3, 4), (5, r) and slope ⁻1

Solution

From slope $m = \dfrac{y_2 - y_1}{x_2 - x_1}$

$^-1 = \dfrac{r - 4}{5 - 3}$

$^-1 = \dfrac{r - 4}{2}$

Cross multiply

$r - 4 = {}^-1 \times 2$

$r - 4 = {}^-2$

$r = {}^-2 + 4$

$r = 2$

(ii) (2, 2+r), (⁻1, 3) and slope *1*

Solution

From slope $m = \dfrac{y_2 - y_1}{x_2 - x_1}$

$$1 = \dfrac{3 - (2 + r)}{^{-}1 - 2}$$

$$1 = \dfrac{3 - 2 - r}{-3}$$

$$1 = \dfrac{1 - r}{^{-}3}$$

Multiply both sides by ⁻3

$$^{-}3 = 1 - r$$

Add *r* to both sides of equation

$$^{-}3 + r = 1 - r + r$$

$$^{-}3 + r = 1$$

Add *3* to both sides

$$^{-}3 + 3 + r = 1 + 3$$

$$r = 4$$

Examples 12.7

(a) Find the equation of the line passing through a point (⁻3, 2) and parallel to a line ⁻4*x* + 5*y* = 25. Leave your answer in a standard form.

Solution

$$^{-}4x + 5y = 25$$

Add 4*x* to both sides

$$^{-}4x + 4x + 5y = 25 + 4x$$

$$5y = 4x + 25$$

Divide both sides by *5*

$$\dfrac{5y}{5} = \dfrac{4x + 25}{5}$$

$$y = \frac{4}{5}x + 5$$

Compare with $y = mx + c$

$$m = \frac{4}{5}$$

For parallel lines $m = m_1 = \frac{4}{5}$

From slope $m = \dfrac{y - y_1}{x - x_1}$

$$\frac{4}{5} = \frac{y - 2}{x - {}^-3}$$

$$\frac{4}{5} = \frac{y - 2}{x + 3}$$

Cross multiply

$$5(y - 2) = 4(x + 3)$$
$$5y - 10 = 4x + 12$$

Add *10* to both sides

$$5y - 10 + 10 = 4x + 12 + 10$$
$$5y = 4x + 22$$

Subtract $4x$ from both sides

$$5y - 4x = 4x + 22 - 4x$$
$$5y - 4x = 22$$
$${}^-4x + 5y = 22$$

Divide both sides by $^-1$

$$\frac{{}^-4x + 5y}{{}^-1} = \frac{22}{{}^-1}$$

$$4x - 5y = {}^-22$$

(b) Find the equation of the line passing through a point
$({}^-3, 8)$ and perpendicular to a line $3x + y = 10$. Give your

answer in a slope intercept form.

Solution

$$3x + y = 10$$

Subtract $3x$ on both sides

$$3x - 3x + y = 10 - 3x$$

$$y = {}^-3x + 10$$

Compare with $y = mx + c$ where m is the slope

$$m = {}^-3$$

For perpendicular lines

$$mm_1 = {}^-1$$

$$m_1 = \frac{{}^-1}{m}$$

$$= \frac{{}^-1}{{}^-3}$$

$$m_1 = \frac{1}{3}$$

From slope $m = \dfrac{y - y_1}{x - x_1}$

$$\frac{1}{3} = \frac{y - 8}{x - {}^-3}$$

$$\frac{1}{3} = \frac{y - 8}{x + 3}$$

Cross multiply

$$3(y - 8) = x + 3$$

Divide both sides by 3

$$\frac{3(y - 8)}{3} = \frac{x + 3}{3}$$

$$y - 8 = \frac{x}{3} + \frac{3}{3}$$

$$y - 8 = \frac{x}{3} + 1$$

Add 8 to both sides of equation

$$y - 8 + 8 = \frac{x}{3} + 1 + 8$$

$$y = \frac{x}{3} + 9$$

Example 12.8

Shade the wanted region on a graph

Given, $2x - y \leq {}^-2$ and

$$x + y \geq 6$$

Solution

Get at least 5 points on a graph using the above two equations. The first 2 points will be the x-intercept $(x_1, 0)$ and the y-intercept $(0, y_1)$ of the first equation then, the second two points will be the x-intercept $(x_2, 0)$ and y-intercept $(0, y_2)$ of the second equation. The 5th point (x_3, y_3) is the intersection of the two lines, and to get this equate the two lines together.

For 1st equation

$$2x - y \leq {}^-2$$

Subtract $2x$ on both sides

$$2x - 2x - y \leq {}^-2 - 2x$$

$${}^-y \leq {}^-2x - 2$$

Divide or multiply both sides by $^-1$

$$\frac{{}^-y}{{}^-1} \geq \frac{-(2x + 2)}{{}^-1}$$

$$y \geq 2x + 2 \qquad\qquad (i)$$

x-intercept $(x_1, 0)$ when $y_1 = 0$

$$y \geq 2x + 2$$
$$0 \geq 2x + 2$$
$$-2 \geq 2x + 2 - 2$$
$$-2 \geq 2x$$

Divide both sides by 2

$$\frac{-2}{2} \geq \frac{2x}{2}$$

$$-1 \geq x$$

$\therefore \quad x_1 = -1 \quad (-1, 0)$

y-intercept $(0, y_1)$ when $x_1 = 0$

$$y \geq 2x + 2$$
$$y \geq 2*0 + 2$$
$$y \geq 2$$

$\therefore \quad y_1 = 2 \quad (0, 2)$

For 2nd equation

$$x + y \geq 6$$

Subtract x on both sides

$$x - x + y \geq 6 - x$$
$$y \geq 6 - x$$
$$y \geq -x + 6 \hspace{4cm} \text{(ii)}$$

x-intercept $(x_2, 0)$ when $y_2 = 0$

$$y \geq -x + 6$$
$$0 \geq -x + 6$$

Add x to both sides

$$x \geq 6$$

$\therefore \quad x_2 = 6 \quad (6, 0)$

y-intercept $(0, y_2)$ when $x_2 = 0$

$$y \geq -x + 6$$
$$y \geq 0 + 6$$
$$y \geq 6$$

$\therefore \quad y_2 = 6 \quad (0, 6)$

From equation *(i)* and *(ii)*

$$y \geq 2x + 2$$

Also $\quad y \geq {}^-x + 6$

$\therefore \quad 2x + 2 = {}^-x + 6$

Collect like terms together

$$2x + x = 6 - 2$$

$$3x = 4$$

Divide both sides by 3

$$\frac{3x}{3} = \frac{4}{3}$$

$$x = \frac{4}{3}$$

Substitute $x = \dfrac{4}{3}$ in equation *(ii)*

$$y \geq {}^-x + 6$$

$$y \geq {}^-\frac{4}{3} + 6$$

$$y \geq \frac{{}^-4 + 18}{3}$$

$$y \geq \frac{14}{3} \qquad \left(\frac{4}{3}, \frac{14}{3}\right)$$

We now have five points $({}^-1, 0)$, $(0, 2)$, $(6, 0)$, $(0, 6)$ and $\left(\dfrac{4}{3}, \dfrac{14}{3}\right)$

Plot the points on the graph and shade the wanted region

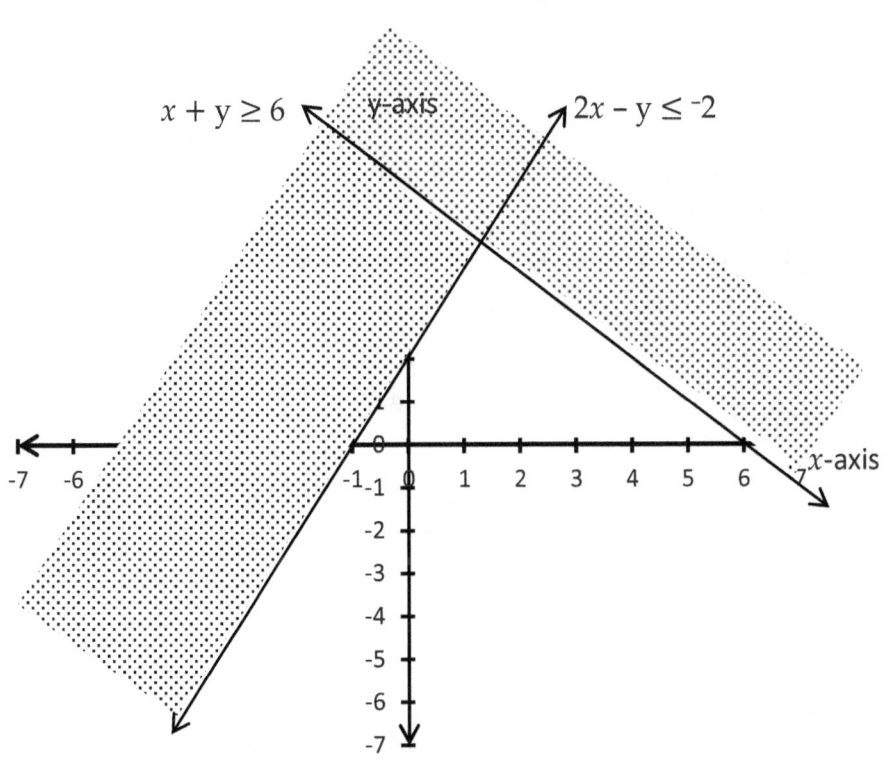

Exercise 12A

(1) In which quadrant do you find the following points:

   K(2, 5), L(⁻1, ⁻3), M(⁻2, 2) and N(3, ⁻4)

(2) Given $A = (0, ⁻2)$, (2, ⁻4), (4, ⁻6)

$B = (1, 4), (2, 5), (1, 3)$
$C = (^-1, 3), (^-2, 5), (^-2, 2)$
$D = (5, ^-1), (^-1, 5)$

(a) Find the Domain and the Range of A, B, C and D
(b) Which of the following are Functions: A, B, C and D
(c) Make a table and a graph for D
(3) Given $f_{(x)} = 3x - 3$ Find the value of $f_{(0)}$ and $f_{(2)}$
(4) The Domain of $f_{(x)} = 3x - 3$ is $D = \{ 1, ^-1, ^-2 \}$
  Find the Range.
(5) Given $g_{(x)} = 3x + 4$. Find $g_{(a-1)}$
(6) Use Function notation equation $x^2 - y = 49$
  to find $f_{(^-5)}$ and $f_{(a)}$
(7) Given the Function $f_{(x)} = x - 2$ and
$$g_{(x)} = x^2$$
Find:

(a) $\dfrac{f(x-h) - f(x)}{h}$

(b) $g_{(x)} + f_{(x)}$
(8) Find the following inverse functions given $f_{(x)} = 22x - 7$
(a) $f_{(x)}^{-1}$
(b) $f o f_{(x)}^{-1}$
(c) $f^{-1} o f_{(x)}$
(9) Given $f_{(x)} = 2x - 3$ and $g_{(x)} = x^2$
  Find (i) $f_{(2)} + g_{(2)}$
    (ii) $f(g_{(2)})$
    (iii) $g(f_{(2)})$
    (iv) $f_{(2)} * g_{(2)}$
(10) Given $f_{(x)} = \dfrac{2}{5}x - 1$.
    Find: (i) $f_{(x)}^{-1}$
      (ii) $f^{-1}(f_{(x)})$
      (iii) $f o f_{(x)}^{-1}$

(11) Graph $f_{(x)} = 2x - 1$ on the coordinate plane

(12) Given $f_{(x)} = \left(\dfrac{1}{8}\right)^x$ and $a > b$ which of the following

must be true:

    A. $f_{(a)} = f_{(b)}$

    B. $f_{(a)} < f_{(b)}$

    C. $f_{(a)} > f_{(b)}$

(13) Write $y = 3x - 3$ in standard form $ax + by = c$

(14) Find the equation of a line in a standard form

    $ax + by = c$ given the slope is $\dfrac{-2}{5}$ and y-intercept is 4

(15) Find the equation of a line in a slope intercept form

    given the slope is 13 and a point on a line is (7, 2)

(16) Given two points (−3, 0) and (3, 8). Find the following:

(i) The distance between the two points

(ii) The midpoint of the line between the two points

(iii) Slope of the line between the two points

(iv) The equation of the line between the two points

(v) The slope of the line parallel to a line between the two points

(vi) The slope of the line perpendicular to a line between The two points

(vii) The x-intercept and y-intercept

(17) Find n in an equation $nx + y = 1$

    which has a solution (1, −1)

(18) Find the equation of a line in standard form with slope 2 and y-intercept −23

(19) Find y given two points and slope of the line

(a) (15, 4), (9, y) and slope −3

(b) (7, 3+y), (10, 3) and slope $-\dfrac{1}{2}$

(20) Find the equation of the line passing through a point (−8, 8) and parallel to a line −3x + 4y = 2. Leave your

answer in a standard form.

(21) Find the equation of the line passing through a point (1, 4) and perpendicular to a line $x + 3y = 6$. Give your answer in a slope intercept form.

(22) Three points on a line are $(6, 3), (5, 0)$ and $(2, n+1)$ Find n.

(23) Find whether the following two lines are parallel, perpendicular or neither

$$x - 2y = {}^-1$$
$$2x - 4y = 3$$

(24) Shade the wanted region on a graph

Given, $4x - y \leq {}^-8$ and

$$x + y \geq 18$$

SOLUTIONS FOR EXERCISE 12A

(1)   K(2, 5) found in 1st quadrant

   L(‾1, ‾3) found in 3rd quadrant

   M(‾2, 2) found in 2nd quadrant

   N(3, ‾4) found in 4th quadrant

(2) $A = (0, ‾2), (2, ‾4), (4, ‾6)$

   $B = (1, 4), (2, 5), (1, 3)$

   $C = (‾1, 3), (‾2, 5), (‾2, 2)$

   $D = (5, ‾1), (‾1, 5)$

(a)   For A   Domain(x) = { 0, 2, 4 }

          Range (y) = { ‾2, ‾4, ‾6 }

   For B   Domain(x) = { 1, 2, 1 }

          Range (y) = { 4, 5, 3 }

   For C   Domain(x) = { ‾1, ‾2, ‾2 }

          Range (y) = { 3, 5, 2 }

   For D   Domain(x) = { 5, ‾1 }

          Range (y) = { ‾1, 5 }

(b) A is a Function because all its Domain or x-values are different from each other.

B is not a Function because it has the same Domain or x-values which are 1.

C is not a Function because it has the same Domain or x-values which are ‾2.

D is a Function because all its Domain or x-values are different from each other.

(c)   For $D = (5, {}^{-}1), ({}^{-}1, 5)$

x	y
5	$^{-}1$
$^{-}1$	5

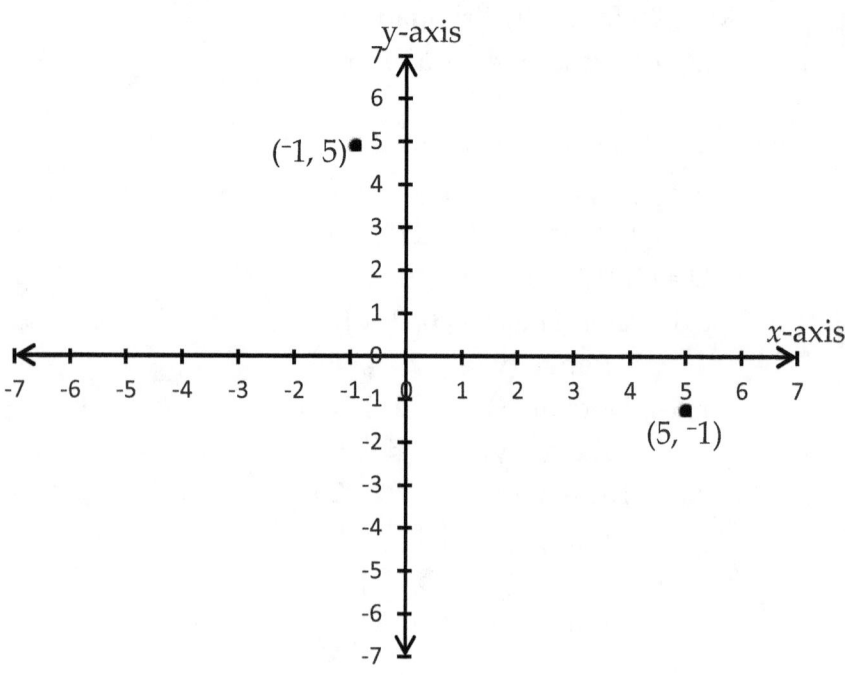

(3) $f_{(x)} = 3x - 3$
$f_{(0)} = 3 \times 0 - 3$
$f_{(0)} = {}^{-}3$
$f_{(2)} = 3 \times 2 - 3$
$f_{(2)} = 6 - 3$
$\quad = 3$

(4) $f_{(x)} = 3x - 3$

$$f_{(1)} = 3(1) - 3$$
$$= 3 - 3$$
$$= 0$$
$$f_{(-1)} = 3(-1) - 3$$
$$= -3 - 3$$
$$= -6$$
$$f_{(-2)} = 3(-2) - 3$$
$$= -6 - 3$$
$$= -9$$
$$\therefore \text{Range} = \{\, 0,\ -6,\ -9\, \}$$

(5)
$$g_{(x)} = 3x + 4.$$
$$g_{(a-1)} = 3(a - 1) + 4$$
$$= 3a - 3 + 4$$
$$g_{(a-1)} = 3a + 1$$

(6)
$$x^2 - y = 49$$

Make y the subject by subtracting x^2 on both sides
$$x^2 - x^2 - y = 49 - x^2$$
$$-y = 49 - x^2$$

Multiply or divide by -1 on both sides
$$y = -49 + x^2$$
$$f_{(x)} = -49 + x^2$$
$$f_{(x)} = x^2 - 49$$
$$f_{(-5)} = -5^2 - 49$$
$$= 25 - 49$$
$$f_{(-5)} = -24$$

Also from $f_{(x)} = x^2 - 49$
$$f_{(a)} = a^2 - 49$$
$$= a^2 - 7^2$$
$$f_{(a)} = (a + 7)(a - 7)$$

7(a) $\quad \dfrac{f_{(x-h)} - f_{(x)}}{h}$

From $f_{(x)} = x - 2$

$\qquad f_{(x-h)} = x - h - 2$

$\therefore \quad \dfrac{f_{(x-h)} - f_{(x)}}{h} = \dfrac{x - h - 2 - (x-2)}{h}$

$\qquad\qquad\qquad = \dfrac{x - h - 2 - x + 2}{h}$

$\qquad\qquad\qquad = \dfrac{^-\cancel{h}}{\cancel{h}}$

$\qquad\qquad\qquad = {}^-1$

(b) $\qquad g_{(x)} + f_{(x)}$

$\qquad\qquad g_{(x)} + f_{(x)} = x^2 + x - 2$

$\qquad\qquad$ Multiply x^2 and $^-2$

$$-2x^2 \underset{\searrow 2x}{\overset{\nearrow {}^-x}{}}$$

a pair or 2 factors whose sum is x

$\qquad\qquad x^2 - x + 2x - 2$

$\qquad\qquad x(x - 1) + 2(x - 1)$

$\qquad\qquad (x - 1)(x + 2)$

$\qquad g_{(x)} + f_{(x)} = (x - 1)(x + 2)$

8(a) $\quad f_{(x)}^{-1}$

$\qquad\qquad f_{(x)} = 22x - 7$

$\qquad\qquad y = 22x - 7$

$\qquad$ Interchange x and y

$$x = 22y - 7$$

Make y the subject by adding 7 on both sides

$$x + 7 = 22y$$

$$22y = x + 7$$

Divide both sides by 22

$$\frac{22y}{22} = \frac{x + 7}{22}$$

$$y = \frac{x + 7}{22}$$

$$\therefore f_{(x)}^{-1} = \frac{x + 7}{22}$$

(b) $$f o f_{(x)}^{-1} = f_{\left(\frac{x+7}{22}\right)}$$

$$= 22\left(\frac{x + 7}{22}\right) - 7$$

$$= x + 7 - 7$$

$$f o f_{(x)}^{-1} = x$$

(c) $$f^{-1} o f_{(x)} = f_{(22x-7)}^{-1}$$

$$= \frac{22x - 7 + 7}{22}$$

$$= \frac{22x}{22}$$

$$f^{-1} o f_{(x)} = x$$

$$\therefore f o f_{(x)}^{-1} = f^{-1} o f_{(x)} = x$$

9(i) $f_{(2)} + g_{(2)}$

$$f_{(x)} = 2x - 3$$
$$f_{(2)} = 2(2) - 3$$
$$= 4 - 3$$
$$= 1$$
$$g_{(x)} = x^2$$
$$g_{(2)} = 2^2$$
$$= 2 \times 2$$
$$= 4$$
$$f_{(2)} + g_{(2)} = 1 + 4$$
$$= 5$$

(ii) $f\left(g_{(2)}\right)$

$$f_{(x)} = 2x - 3 \qquad g_{(x)} = x^2$$
$$f\left(g_{(x)}\right) = f_{(x^2)}$$
$$= 2x^2 - 3$$
$$f\left(g_{(2)}\right) = 2(2)^2 - 3$$
$$= 2*4 - 3$$
$$= 8 - 3$$
$$= 5$$

(iii) $g\left(f_{(2)}\right)$

$$g\left(f_{(x)}\right) = g_{(2x - 3)}$$
$$= (2x - 3)^2$$
$$g\left(f_{(2)}\right) = [2(2) - 3]^2$$
$$= (4 - 3)^2$$
$$= 1$$

(iv) $f_{(2)} * g_{(2)}$

$$f_{(x)} = 2x - 3$$
$$f_{(2)} = 2(2) - 3$$
$$= 4 - 3$$
$$= 1$$

$$g_{(x)} = x^2$$
$$g_{(2)} = 2^2$$
$$= 2 \times 2$$
$$= 4$$
$$f_{(2)} * g_{(2)} = 1 \times 4$$
$$= 4$$

10(i) $f_{(x)}^{-1}$

$$f_{(x)} = \frac{2}{5}x - 1$$

$$y = \frac{2}{5}x - 1$$

Interchange x and y

$$x = \frac{2}{5}y - 1$$

Make y the subject by adding 1 on both sides

$$\frac{2}{5}y = x + 1$$

Multiply a reciprocal of coefficient of y on both sides

$$\frac{5}{2} \times \frac{2}{5}y = \frac{5}{2}(x + 1)$$

$$y = \frac{5}{2}x + \frac{5}{2}$$

$$\therefore f_{(x)}^{-1} = \frac{5}{2}x + \frac{5}{2}$$

(ii) $$f^{-1}(f_{(x)}) = f_{(\frac{2}{5}x - 1)}^{-1}$$

$$= \frac{5}{2}\left(\frac{2}{5}x - 1\right) + \frac{5}{2}$$

$$= x - \frac{5}{2} + \frac{5}{2}$$
$$= x$$

(iii) $\qquad f \circ f^{-1}_{(x)} = f_{\left(\frac{5}{2}x + \frac{5}{2}\right)}$

$$= \frac{2}{5}\left(\frac{5}{2}x + \frac{5}{2}\right) - 1$$
$$= x + 1 - 1$$
$$= x$$
$$\therefore \quad f \circ f^{-1}_{(x)} = f^{-1} \circ f_{(x)} = x$$

(11) $\qquad f_{(x)} = 2x - 1$
$$y = 2x - 1$$
when $x = 0$
$$y = 2(0) - 1$$
$$= 0 - 1$$
$$= {}^-1 \qquad \text{one of the point is } (0, {}^-1)$$
when $y = 0$
$$0 = 2x - 1$$
$$2x = 1$$
$$x = \frac{1}{2} \qquad \text{another point is } \left(\frac{1}{2}, 0\right)$$

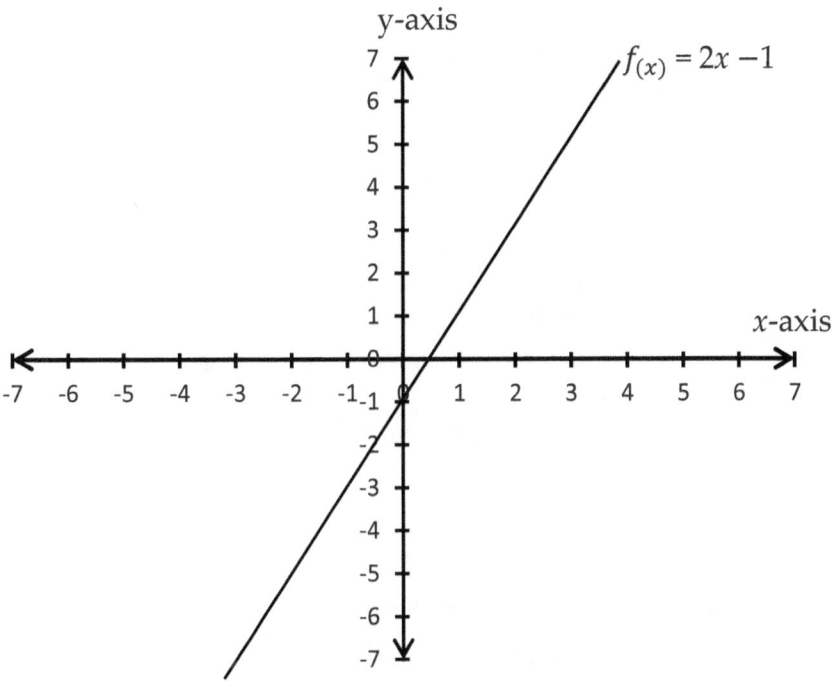

(12)
$$f_{(x)} = \left(\frac{1}{8}\right)^x$$

$$f_{(a)} = \left(\frac{1}{8}\right)^a$$

$$f_{(b)} = \left(\frac{1}{8}\right)^b$$

Since $a > b$, the answer is B. $f_{(a)} < f_{(b)}$

(13)
$$y = 3x - 3$$
Subtract $3x$ on both sides
$$^-3x + y = 3x - 3x - 3$$
$$^-3x + y = {}^-3$$

Multiply by ⁻1 on both sides
$$3x - y = 3$$

(14)   From $y = mx + c$
$$y = \frac{-2}{5}x + 4$$

Add $\frac{2}{5}x$ to both sides

$$\frac{2}{5}x + y = \frac{-2}{5}x + \frac{2}{5}x + 4$$

$$\frac{2}{5}x + y = 4$$

Multiply both sides by 5
$$5\left(\frac{2}{5}x + y\right) = 4 \times 5$$
$$2x + 5y = 20$$

(15)   From slope $m = \dfrac{y - y_1}{x - x_1}$

$$13 = \frac{y - 2}{x - 7}$$

Cross multiply
$$y - 2 = 13(x - 7)$$
$$y - 2 = 13x - 91$$
Add 2 to both sides
$$y - 2 + 2 = 13x - 91 + 2$$
$$y = 13x - 89$$

(16)   Points (⁻3, 0) and (3, 8)

(i)   From
$$d = \sqrt{(x_1 - x_2)^2 + (y_1 - y_2)^2}$$
$$= \sqrt{(^-3 - 3)^2 + (0 - 8)^2}$$
$$= \sqrt{^-6^2 + ^-8^2}$$

$$= \sqrt{36 + 64}$$
$$= \sqrt{100}$$
$$= \sqrt{10^2}$$
$$= 10$$

∴   the distance between the two points is *10*

(ii)   Using the midpoint formula

$$= \left(\frac{x_1 + x_2}{2} , \frac{y_1 + y_2}{2} \right)$$

$$= \left(\frac{^-3 + 3}{2} , \frac{0 + 8}{2} \right)$$

$$= \left(\frac{0}{2} , \frac{8}{2} \right)$$

Midpoint $= (0, 4)$

(iii)   From slope $m = \dfrac{y_2 - y_1}{x_2 - x_1}$

$$= \frac{8 - 0}{3 - ^-3}$$

$$= \frac{8}{3 + 3}$$

$$= \frac{8}{6}$$

$$m = \frac{4}{3}$$

(iv)   From $m = \dfrac{y - y_1}{x - x_1}$

$$\frac{4}{3} = \frac{y - 0}{x - {}^-3}$$

$$\frac{4}{3} = \frac{y}{x + 3}$$

Cross multiply

$$3y = 4(x + 3)$$

Divide both sides by 3

$$\frac{3y}{3} = \frac{4}{3}(x + 3)$$

$$y = \frac{4}{3}(x + 3)$$

$$y = \frac{4}{3}x + 4$$

that's the equation of the line between the two points

(v)   For parallel lines

$$m = m_1 = \frac{4}{3}$$

(vi)   For perpendicular lines

$$m_1 m_2 = {}^-1$$

$$\frac{4}{3} m_2 = {}^-1$$

Multiply both sides by 3

$$3 \times \frac{4}{3} m_2 = {}^-1 \times 3$$

$$4m_2 = {}^-3$$

Divide both sides by 4

$$\frac{4m_2}{4} = \frac{{}^-3}{4}$$

$$m_2 = \frac{^-3}{4}$$

∴   the slope of the line perpendicular to a line

between the two points is $\dfrac{^-3}{4}$

(vii)   x-intercept is the point on the line where $y = 0$

From the equation of the line

$$y = \frac{4}{3}x + 4$$

$$0 = \frac{4}{3}x + 4$$

Subtract 4 from both sides

$$0 - 4 = \frac{4}{3}x + 4 - 4$$

$$^-4 = \frac{4}{3}x$$

Multiply both sides by $\dfrac{3}{4}$

$$^-4 \times \frac{3}{4} = \frac{3}{4} \times \frac{4}{3}x$$

$$^-3 = x$$

∴   x-intercept is ($^-3$, 0)

y-intercept is a point on the line where $x = 0$

Also from the equation of the line

$$y = \frac{4}{3}x + 4$$

$$y = \frac{4}{3} \times 0 + 4$$

$$y = 0 + 4$$

$$y = 4$$

$\therefore$ y-intercept is $(0, 4)$

OR   compare the equation of the line

$$y = \frac{4}{3}x + 4 \quad \text{with}$$

$$y = mx + c$$

$c = 4$, at y-intercept $(0, 4)$

(17)   From a solution $(1, {}^-1)$ $x = 1$, $y = {}^-1$

$$nx + y = 1$$

$$n(1) + {}^-1 = 1$$

$$n - 1 = 1$$

$$n = 1 + 1$$

$$n = 2$$

(18)   From $y = mx + c$ $m = 2$, $c = {}^-23$

$$y = 2x + {}^-23$$

$$y = 2x - 23$$

$${}^-2x + y = {}^-23$$

$$2x - y = 23$$

19(a) $(15, 4)$, $(9, y)$ and slope ${}^-3$

From slope $m = \dfrac{y_2 - y_1}{x_2 - x_1}$

$${}^-3 = \frac{y - 4}{9 - 15}$$

$${}^-3 = \frac{y - 4}{{}^-6}$$

Cross multiply

$$y - 4 = {}^-3 \times {}^-6$$

$$y - 4 = 18$$

$$y = 18 + 4$$

$$y = 22$$

(b)   (7, 3+y), (10, 3) and slope $-\dfrac{1}{2}$

From slope $m = \dfrac{y_2 - y_1}{x_2 - x_1}$

$$-\frac{1}{2} = \frac{3 - (3 + y)}{10 - 7}$$

$$-\frac{1}{2} = \frac{3 - 3 - y}{3}$$

$$-\frac{1}{2} = \frac{-y}{3}$$

Cross multiply

$$^-2y = {}^-3$$

Divide both sides by $^-2$

$$\frac{^-2y}{^-2} = \frac{^-3}{^-2}$$

$$y = \frac{3}{2}$$

(20) $(^-8, 8)$ and parallel to a line $^-3x + 4y = 2$

$$^-3x + 4y = 2$$

Add $3x$ to both sides

$$^-3x + 3x + 4y = 2 + 3x$$

$$4y = 3x + 2$$

Divide both sides by 4

$$\frac{4y}{4} = \frac{3x + 2}{4}$$

$$\frac{4y}{4} = \frac{3}{4}x + \frac{2}{4}$$

$$y = \frac{3}{4}x + \frac{1}{2}$$

Compare with $y = m_1x + c$

$$m_1 = \frac{3}{4}$$

For parallel lines $m = m_1 = \frac{3}{4}$

From slope $m = \dfrac{y - y_1}{x - x_1}$

$$\frac{3}{4} = \frac{y - 8}{x - \,^-8}$$

$$\frac{3}{4} = \frac{y - 8}{x + 8}$$

Cross multiply

$4(y - 8) = 3(x + 8)$

$4y - 32 = 3x + 24$

Add 32 to both sides

$4y - 32 + 32 = 3x + 24 + 32$

$4y = 3x + 56$

Subtract $3x$ on both sides

$4y - 3x = 3x - 3x + 56$

$4y - 3x = 56$

Multiply both sides by $^-1$

$3x - 4y = \,^-56$

(21) $(1, 4)$ and perpendicular to a line $x + 3y = 6$

$x + 3y = 6$

Subtract x on both sides

$x - x + 3y = 6 - x$

$3y = \,^-x + 6$

Divide by 3 on each term

$$\frac{3y}{3} = \frac{^-x}{3} + \frac{6}{3}$$

$$y = -\frac{1}{3}x + 2$$

Compare with $y = mx + c$ where m is the slope

$$m = -\frac{1}{3}$$

For perpendicular lines

$$mm_1 = -1$$

$$-\frac{1}{3}m_1 = -1$$

Multiply both sides by -3

$$-3 \times -\frac{1}{3}m_1 = -1 \times -3$$

$$m_1 = 3$$

From slope $m = \dfrac{y - y_1}{x - x_1}$

$$3 = \frac{y - 4}{x - 1}$$

Cross multiply

$$y - 4 = 3(x - 1)$$
$$y - 4 = 3x - 3$$

Add 4 to both sides

$$y - 4 + 4 = 3x - 3 + 4$$
$$y = 3x + 1$$

(22) $(6, 3), (5, 0)$ and $(2, n+1)$

From slope $m = \dfrac{y_2 - y_1}{x_2 - x_1}$

$$m = \frac{0 - 3}{5 - 6}$$

$$m = \frac{-3}{-1}$$

$$m = 3$$

Also from slope $m = \dfrac{y_2 - y_1}{x_2 - x_1}$

$$3 = \dfrac{n+1-0}{2-5}$$

$$3 = \dfrac{n+1}{^-3}$$

Multiply both sides by $^-3$

$$^-3 \times 3 = \dfrac{n+1}{^-3} \times {^-3}$$

$$^-9 = n + 1$$

$$n + 1 = {^-9}$$

Subtract 1 on both sides

$$n + 1 - 1 = {^-9} - 1$$

$$n = {^-10}$$

(23) Rewrite the two lines into slope-intercept form

$$x - 2y = {^-1}$$

$$x - x - 2y = {^-x} - 1$$

$$-2y = {^-x} - 1$$

$$\dfrac{-2y}{-2} = \dfrac{^-x}{-2} - \dfrac{1}{-2}$$

$$y = \dfrac{1}{2}x + \dfrac{1}{2}$$

Compare with $y = m_1 x + c$

$$m_1 = \dfrac{1}{2}$$

For 2nd line

$$2x - 4y = 3$$

$$2x - 2x - 4y = {^-2x} + 3$$

$$^-4y = {^-2x} + 3$$

$$\frac{^-4y}{^-4} = \frac{^-2x}{^-4} + \frac{3}{-4}$$

$$y = \frac{1}{2}x - \frac{3}{4}$$

Compare with $y = m_2x + c$

$$m_2 = \frac{1}{2}$$

Since $m_1 = m_2 = \dfrac{1}{2}$

∴ the two lines are parallel

(24) Given, $4x - y \le {}^-8$ and

$x + y \ge 18$

For 1st equation

$4x - y \le {}^-8$

Subtract $4x$ on both sides

$4x - 4x - y \le {}^-8 - 4x$

$^-y \le {}^-4x - 8$

Divide or multiply both sides by $^-1$

$$\frac{^-y}{^-1} \ge \frac{-(4x + 8)}{^-1}$$

$y \ge 4x + 8$ (i)

x-intercept $(x_1, 0)$ when $y_1 = 0$

$y \ge 4x + 8$

$0 \ge 4x + 8$

$^-8 \ge 4x + 8 - 8$

$^-8 \ge 4x$

Divide both sides by 4

$$\frac{^-8}{4} \ge \frac{4x}{4}$$

$^-2 \ge x$

∴ $x_1 = {}^-2$ $({}^-2, 0)$

y-intercept $(0, y_1)$ when $x_1 = 0$

$y \ge 4x + 8$

$$y \geq 4*0 + 8$$
$$y \geq 8$$
$$\therefore y_1 = 8 \quad (0, 8)$$

For 2nd equation

$$x + y \geq 18$$

Subtract x on both sides

$$x - x + y \geq 18 - x$$
$$y \geq 18 - x$$
$$y \geq {}^-x + 18 \qquad\qquad\qquad (ii)$$

x-intercept $(x_2, 0)$ when $y_2 = 0$

$$y \geq {}^-x + 18$$
$$0 \geq {}^-x + 18$$

Add x to both sides

$$x \geq 18$$
$$\therefore x_2 = 18 \quad (18, 0)$$

y-intercept $(0, y_2)$ when $x_2 = 0$

$$y \geq {}^-x + 18$$
$$y \geq 0 + 18$$
$$y \geq 18$$
$$\therefore y_2 = 18 \quad (0, 18)$$

From equation *(i)* and *(ii)*

$$y \geq 4x + 8$$
$$y \geq {}^-x + 18$$
$$\therefore 4x + 8 = {}^-x + 18$$

Collect like terms together

$$4x + x = 18 - 8$$
$$5x = 10$$

Divide both sides by 5

$$\frac{\cancel{5}x}{\cancel{5}} = \frac{\cancel{10}^2}{\cancel{5}}$$
$$x = 2$$

Substitute $x = 2$ in equation *(ii)*

$$y \geq {}^-x + 18$$

$$y \geq {}^-2 + 18$$
$$y \geq 16 \qquad (2, 16)$$

We now have five points $(^-2, 0)$, $(0, 8)$, $(18, 0)$, $(0, 18)$ and $(2, 16)$

Plot the points on the graph and shade the wanted region.

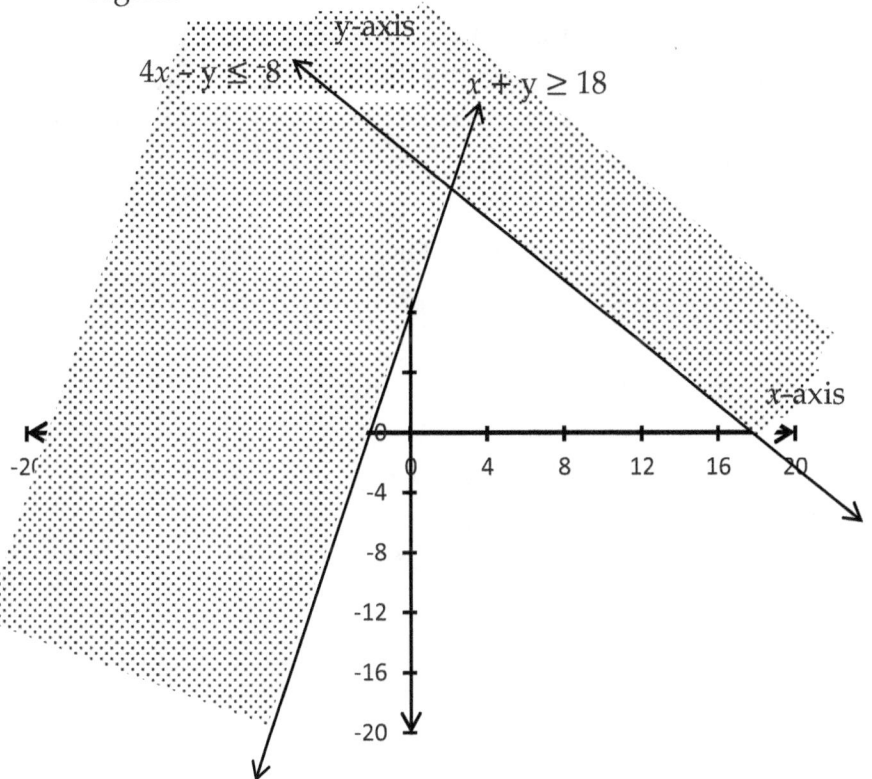

Exercise 12B

(1) In which quadrant do you find the following points?
$\qquad$ K(2, ⁻5), L(⁻1, 3), M(⁻2, ⁻2) and N(3, 4)

(2) Given A = (1, 0), (3, ⁻2), (5, ⁻4)
$\qquad$ B = (7, 5), (2, 5), (1, 3)
$\qquad$ C = (⁻1, 4), (0, 5), (⁻1, 2)
$\qquad$ D = (6, 4), (4, 4)

(a) Find the Domain and the Range of A, B, C and D

(b) Which of the following are Functions: A, B, C and D.

(c) Make a table and a graph for D

(3) Given $f_{(x)} = 2x - 4$. Find the value of $f_{(-1)}$ and $f_{(3)}$

(4) The domain of $f_{(x)} = x - 2$ is $D = \{ 1, 2, 3 \}$
$\qquad$ Find the range.

(5) Given $g_{(x)} = 2x + 2$. Find $g_{(a-1)}$

(6) Use Function notation equation $3x - y = 21$
$\qquad$ to find $f_{(9)}$ and $f_{(a +7)}$

(7) Given the Function $f_{(x)} = x - 4$ and
$$g_{(x)} = x^2$$
$\qquad$ Find:

(a) $\dfrac{f(x-h) - f(x)}{h}$

(b) $g_{(x)} + 2f_{(x)}$

(8) Find the following inverse functions given $f_{(x)} = 11x - 6$

(a) $f_{(x)}^{-1}$

(b) $f \circ f_{(x)}^{-1}$

(c) $f^{-1} \circ f_{(x)}$

(9) Given $f_{(x)} = 3x - 4$ and $g_{(x)} = x^2$
$\qquad$ Find (i) $f_{(3)} + g_{(2)}$
$\qquad\qquad$ (ii) $f\left(g_{(2)}\right)$
$\qquad\qquad$ (iii) $g\left(f_{(3)}\right)$
$\qquad\qquad$ (iv) $f_{(3)} * g_{(2)}$

(10) Given $f_{(x)} = \dfrac{3}{4}x - 2$.

Find: (i) $f_{(x)}^{-1}$

(ii) $f^{-1}(f_{(x)})$

(iii) $f \circ f_{(x)}^{-1}$

(11) Graph $f_{(x)} = 6x - 6$ on the coordinate plane

(12) Given $f_{(x)} = \left(\dfrac{1}{10}\right)^x$ and $\dfrac{a}{b} < 1$ which of the following

must be true:

A. $f_{(a)} = f_{(b)}$

B. $f_{(a)} < f_{(b)}$

C. $f_{(a)} > f_{(b)}$

(13) Write $y = 9x - 17$ in standard form $ax + by = c$

(14) Find the equation of a line in a standard form

$ax + by = c$, given the slope is $\dfrac{-1}{3}$ and y-intercept is 8

(15) Find the equation of a line in a slope intercept form
given the slope is 15 and a point on a line is (-3, -4)

(16) Given two points (-2, 0) and (1, 4). Find the following:

(i) The distance between the two points

(ii) The midpoint of the line between the two points

(iii) Slope of the line between the two points

(iv) The equation of the line between the two points

(v) The slope of the line parallel to a line between the two
points

(vi) The slope of the line perpendicular to a line between
the two points

(vii) The x-intercept and y-intercept

(17) Find n in an equation $nx + 2y = 1$
which has a solution (2, -1)

(18) Find the equation of a line in standard form with slope
7 and y-intercept -43

(19) Find y given two points and slope of the line

(a) (3, 5), (‾2, y) and slope ‾*11*

(b) (‾2, 1+y), (6, 2) and slope $-\dfrac{7}{8}$

(20) Find the equation of the line passing through a point (‾7, ‾9) and parallel to a line ‾$2x + 3y = ‾6$. Leave your answer in a standard form.

(21) Find the equation of the line passing through a point (8, 5) and perpendicular to a line $2x + 4y = 5$. Give your answer in a slope intercept form.

(22) Three points on a line are (*1, 3*), (*4, 4*) and (‾*5, n+2*) Find *n*.

(23) Find whether the following two lines are parallel, perpendicular or neither

$$3x - 27y = ‾13$$
$$18x + 2y = 17$$

(24) Shade the wanted region on a graph

Given, $4x - y \le ‾8$ and

$$x + y \ge 18$$

CHAPTER 13

LOGARITHM

From powers or <u>exponential function</u> like $a^m = n$ (seen in chapter 3), we can express m in terms of a and n by introducing a form of <u>logarithm</u> and it becomes $m = \log_a n$, it is read as m is the logarithm of n to the base a. Also a function $m = \log_a n$ is called a <u>logarithmic function</u>.

Rules of Logarithm
Examples 1
(a) $\qquad n = \log_a m \qquad$ means $\quad m = a^n$

$\qquad$ Find $x \qquad\quad 2 = \log_{10} x$

Solution

$$2 = \log_{10} x$$
$$x = 10^2$$
$$= 10 \times 10$$
$$x = 100$$

(b) $\qquad \log_a a = 1 \quad$ and $\quad \log_a 1 = 0$

$\qquad\quad \log_2 2 = 1 \quad$ and $\quad \log_2 1 = 0$

(c) $\qquad \log_a mn = \log_a m + \log_a n$

$\qquad$ Simplify $\quad \log_{10} 2 + \log_{10} 5$

Solution

$$\log_{10} 2 + \log_{10} 5 = \log_{10}(2 \times 5)$$
$$= \log_{10} 10$$
$$= 1$$

(d) $\qquad \log_a {}^m/_n = \log_a m - \log_a n$

Simplify $\quad \log_{10} 6 - \log_{10} 3$

Solution

$$\log_{10} 6 - \log_{10} 3 = \log_{10}\left({}^6/_3\right)$$
$$= \log_{10} 2$$

e(i) $\qquad \log_a m^r = r \log_a m$

Find r $\quad \log_{10} 1000^r = 6$

Solution

$$\log_{10} 1000^r = 6$$
$$r \log_{10} 1000 = 6$$
$$\text{Let} \quad \log_{10} 1000 = y$$
$$10^y = 1000$$
$$10^y = 10 \times 10 \times 10$$
$$10^y = 10^3$$
$$y = 3$$
$$r \log_{10} 1000 = 6$$
$$r * 3 = 6$$

Divide both sides by 3
$$\frac{r * 3}{3} = \frac{6^2}{3}$$
$$r = 2$$

OR $\quad \log_{10} 1000^r = 6$
$$r \log_{10} 1000 = 6$$
$$r \log_{10} 10^3 = 6$$
$$3r \log_{10} 10 = 6$$
$$\text{But} \quad \log_{10} 10 = 1$$
$$3r * 1 = 6$$

Divide both sides by 3
$$\frac{3r}{3} = \frac{6^2}{3}$$
$$r = 2$$

(ii) Solve $3^{6-3x} = \dfrac{1}{27}$

Solution

$$3^{6-3x} = \dfrac{1}{27}$$

$$3^{6-3x} = \dfrac{1}{3^3}$$

$$3^{6-3x} = 3^{-3}$$

Introduce log_3 on both sides

$$\log_3 3^{6-3x} = \log_3 3^{-3}$$

$$(6-3x)\log_3 3 = {}^-3\log_3 3$$

But $\log_3 3 = 1$

$$(6-3x) * 1 = {}^-3 * 1$$

$$6 - 3x = {}^-3$$

Subtract 6 on both sides

$$6 - 6 - 3x = {}^-3 - 6$$

$${}^-3x = {}^-9$$

Divide by $^-3$ on both sides

$$\dfrac{^-3x}{^-3} = \dfrac{^-9^3}{^-3}$$

$$x = 3$$

(iii) Simplify $3 + \log_2 x^3$

Solution

$$3 + \log_2 x^3 = 3 + 3\log_2 x$$
$$= 3(\,1 + \log_2 x\,)$$

From $\log_a a = 1$,

$$\log_2 2 = 1$$

$$3(\,1 + \log_2 x\,) = 3(\,\log_2 2 + \log_2 x\,)$$

From

$$\log_a m + \log_a n = \log_a mn$$
$$3(\,\log_2 2 + \log_2 x\,) = 3\log_2 2x$$

(f) $\qquad \log_a a^x = x$

Simplify $\quad \log_5 5^9$

Solution

$$\log_5 5^9 = 9 \log_5 5$$
$$= 9 * 1$$
$$\therefore \ \log_5 5^9 = 9$$

(g) $\qquad a^{\log_a x} = x$

Simplify $\quad 3^{\log_3 7}$

Solution

$$\text{Let} \quad y = 3^{\log_3 7}$$

Introduce $\log_3$ on both sides

$$\log_3 y = \log_3 3^{\log_3 7}$$
$$\log_3 y = \log_3 7 * \log_3 3$$

But $\quad \log_3 3 = 1$

$$\log_3 y = \log_3 7 * 1$$
$$\log_3 y = \log_3 7$$
$$y = 7$$
$$\therefore \quad 3^{\log_3 7} = 7$$

(h) **Change of base rule**

$$\log_a m = \frac{\log_b m}{\log_b a}$$

Convert $\quad \log_3 2$ to base 10

$$\log_3 2 = \frac{\log_{10} 2}{\log_{10} 3}$$

Exercise 13A

(1) Find x $\quad 5 = \log_2 x$

(2) Find x $\quad 3 = \log_{10} x$

(3) Simplify $\log_5 2 + \log_5 6$

(4) Simplify $\log_2 5 + \log_2 7$

(5) Simplify $\log_{10} 81 - \log_{10} 3$

(6) Simplify $\log_3 81 - \log_3 9$

(7) Find r $\quad \log_3 243^r = 30$

(8) Find x $\quad \log_7 343^{2x} = 54$

(9) Solve $\quad 8^{10-2x} = \dfrac{1}{64}$

(10) Solve $\quad 4^{6-3x} = \dfrac{1}{64}$

(11) Simplify $2 + \log_{10} y^2$

(12) Simplify $9^{\log_9 77}$

(13) Convert $\log_{25} 3$ to base 5

(14) Convert $\log_{16} 4$ to base 2

(15) Solve for x $\quad 2^{x+2} = 4^{x-8}$

SOLUTIONS FOR EXERCISE 13A

(1)
$$5 = \log_2 x$$
$$x = 2^5$$
$$= 2 \times 2 \times 2 \times 2 \times 2$$
$$x = 32$$

(2)
$$3 = \log_{10} x$$
$$x = 10^3$$
$$= 10 \times 10 \times 10$$
$$x = 1000$$

(3)
$$\log_5 2 + \log_5 6 = \log_5 (2 \times 6)$$
$$= \log_5 12$$

(4)
$$\log_2 5 + \log_2 7 = \log_2 (5 \times 7)$$
$$= \log_2 35$$

(5)
$$\log_{10} 81 - \log_{10} 3 = \log_{10}\left(^{81}/_3\right)$$
$$= \log_{10} 27$$

(6)
$$\log_3 81 - \log_3 9 = \log_3\left(^{81}/_9\right)$$
$$= \log_3 9$$
$$= \log_3 3^2$$
$$= 2\log_3 3$$
$$= 2$$

(7)
$$\log_3 243^r = 30$$
$$r \log_3 243 = 30$$
$$\text{Let } x = \log_3 243$$

288

$$3^x = 243$$
$$3^x = 3 \times 3 \times 3 \times 3 \times 3$$
$$3^x = 3^5$$
$$x = 5$$
$$r \log_3 243 = 30$$
$$r * x = 30$$
$$r * 5 = 30$$

Divide both sides by 5

$$\frac{r * \cancel{5}}{\cancel{5}} = \frac{\cancel{30}^6}{\cancel{5}}$$
$$r = 6$$

OR $\quad \log_3 243^r = 30$
$$r \log_3 243 = 30$$
$$r \log_3 3^5 = 30$$
$$5r \log_3 3 = 30$$

But $\quad \log_3 3 = 1$
$$5r * 1 = 30$$

Divide both sides by 3

$$\frac{5r}{\cancel{5}} = \frac{\cancel{30}^6}{\cancel{5}}$$
$$r = 6$$

(8) $\quad \log_7 343^{2x} = 54$
$$2x \log_7 343 = 54$$

Let $\quad p = \log_7 343$
$$7^p = 343$$
$$7^p = 7 \times 7 \times 7$$
$$7^p = 7^3$$
$$p = 3$$
$$2x \log_7 343 = 54$$
$$2x * p = 54$$
$$2x * 3 = 54$$

Divide both sides by 6

$$\frac{x * \cancel{6}}{\cancel{6}} = \frac{\cancel{54}^{9}}{\cancel{6}}$$

$$x = 9$$

OR $\log_7 343^{2x} = 54$

$2x \log_7 343 = 54$

$2x \log_7 7^3 = 54$

$6x \log_7 7 = 54$

But $\log_7 7 = 1$

$6x * 1 = 54$

Divide both sides by 6

$$\frac{\cancel{6}x}{\cancel{6}} = \frac{\cancel{54}^{9}}{\cancel{6}}$$

$$x = 9$$

(9) $$8^{10 - 2x} = \frac{1}{64}$$

$$8^{10 - 2x} = \frac{1}{8^2}$$

$$8^{10 - 2x} = 8^{-2}$$

Introduce $\log_8$ on both sides

$\log_8 8^{10 - 2x} = \log_8 8^{-2}$

$(10 - 2x) \log_8 8 = {}^-2 \log_8 8$

But $\log_8 8 = 1$

$(10 - 2x) * 1 = {}^-2 * 1$

$10 - 2x = {}^-2$

Subtract 10 on both sides

$10 - 10 - 2x = {}^-2 - 10$

$^-2x = {}^-12$

Divide by $^-2$ on both sides

290

$$\frac{\bcancel{-2}x}{\bcancel{-2}} = \frac{\bcancel{-12}^6}{\bcancel{-2}}$$

$$x = 6$$

(10) $\qquad 4^{6-3x} = \dfrac{1}{64}$

$$4^{6-3x} = \dfrac{1}{4^3}$$

$$4^{6-3x} = 4^{-3}$$

Introduce log_4 on both sides

$$\log_4 4^{6-3x} = \log_4 4^{-3}$$

$$(6 - 3x)\log_4 4 = {}^-3 \log_4 4$$

But $\quad \log_4 4 = 1$

$$(6 - 3x) * 1 = {}^-3 * 1$$

$$6 - 3x = {}^-3$$

Subtract 6 on both sides

$$6 - 6 - 3x = {}^-3 - 6$$

$${}^-3x = {}^-9$$

Divide both sides by ⁻3

$$\frac{{}^-\bcancel{3}x}{{}^-\bcancel{3}} = \frac{{}^-\bcancel{9}^3}{{}^-\bcancel{3}}$$

$$x = 3$$

(11) $\qquad 2 + \log_{10} y^2 = 2 + 2\log_{10} y$

$$= 2(1 + \log_{10} y)$$

From $\quad \log_a a = 1,$

$$\log_{10} 10 = 1$$

$$2(1 + \log_{10} y) = 2(\log_{10} 10 + \log_{10} y)$$

From

$$\log_a m + \log_a n = \log_a mn$$

$$2(\log_{10} 10 + \log_{10} y) = 2\log_{10} 10y$$

(12) $\qquad$ $9^{\log_9 77}$

Let $x = 9^{\log_9 77}$

Introduce $\log_9$ on both sides

$$\log_9 x = \log_9 9^{\log_9 77}$$

$$\log_9 x = \log_9 77 * \log_9 9$$

But $\log_9 9 = 1$

$$\log_9 x = \log_9 77 * 1$$

$$\log_9 x = \log_9 77$$

$$x = 77$$

$$\therefore \quad 9^{\log_9 77} = 77$$

(13) $\qquad$ $\log_{25} 3$ to base 5

From $\log_a m = \dfrac{\log_b m}{\log_b a}$

$$\log_{25} 3 = \dfrac{\log_5 3}{\log_5 25}$$

$$= \dfrac{\log_5 3}{\log_5 5^2}$$

$$= \dfrac{\log_5 3}{2 \log_5 5}$$

$$= \dfrac{\log_5 3}{2}$$

(14) $\qquad$ $\log_{16} 4$ to base 2

From $\log_a m = \dfrac{\log_b m}{\log_b a}$

$$\log_{16} 4 = \dfrac{\log_2 4}{\log_2 16}$$

$$= \frac{\log_2 2^2}{\log_2 2^4}$$

$$= \frac{2\log_2 2}{4\log_2 2}$$

$$= \frac{2}{4}$$

$$= \frac{1}{2}$$

(15) $\qquad 2^{x+2} = 4^{x-8}$

$\qquad\qquad 2^{x+2} = 2^{2(x-8)}$

Introduce log_2 on both sides

$\qquad \log_2 2^{x+2} = \log_2 2^{2(x-8)}$

$(x+2)\log_2 2 = 2(x-8)\log_2 2$

But $\quad \log_2 2 = 1$

$(x+2) * 1 = 2(x-8) * 1$

$\qquad x + 2 = 2x - 16$

Collect like terms together

$\qquad x - 2x = {}^-16 - 2$

$\qquad\qquad {}^-x = {}^-18$

Divide both sides by $^-1$

$$\frac{{}^-x}{{}^-1} = \frac{{}^-18}{{}^-1}$$

$\qquad\qquad x = 18$

Exercise 13B

(1) Find x $4 = \log_3 x$

(2) Find x $3 = \log_5 x$

(3) Simplify $\log_9 12 + \log_9 2$

(4) Simplify $\log_2 6 + \log_2 7$

(5) Simplify $\log_{10} 102 - \log_{10} 3$

(6) Simplify $\log_3 121 - \log_3 11$

(7) Find p $\log_6 216^p = 30$

(8) Find n $\log_{10} 1000^{2n} = 78$

(9) Solve $7^{22 - 3x} = \dfrac{1}{49}$

(10) Solve $3^{26 + 5x} = \dfrac{1}{81}$

(11) Simplify $3 + \log_5 p^3$

(12) Simplify $5^{\log_5 70}$

(13) Convert $\log_{16} 8$ to base 2

(14) Convert $\log_{64} 6$ to base 4

(15) Solve for x $6^{x + 6} = 36^{x - 7}$

CHAPTER 14

CONVERTING U.S AND METRIC UNITS

Table 14.1

Kilo	Hecto	Deka	deci	centi	milli
1000	100	10	0.1	0.01	0.001

U.S to Metric	**Metric to U.S**
1 mile ≈ 1.61 kilometers	1 kilometer ≈ 0.62 mile
1 foot ≈ 0.31 meter	1 meter ≈ 3.3 feet
1 yard ≈ 0.91 meter	1 meter ≈ 1.09 yards
1 inch ≈ 2.54 centimeters	1 centimeter ≈ 0.39 inch
1 pound ≈ 0.45 kilogram	1 kilogram ≈ 2.2 pounds
1 ounce ≈ 28.35 grams	1 gram ≈ 0.035 ounce
1 gallon ≈ 3.78 liters	1 liter ≈ 0.26 gallon
1 quart ≈ 0.95 liter	1 liter ≈ 1.06 quarts
1 pint ≈ 0.47 liter	1 liter ≈ 2.11 pints
1 fluid ounce ≈ 0.03 liter	1 liter ≈ 33.78 fluid ounces

$$\textbf{Formula for unit ratio} = \frac{\textit{New unit}}{\textit{original unit}}$$

14-1 Convert units using table *14.1*
Examples 14.1

Convert *3000* meters to kilometers

Solution

$$1 \; km = 1000 \; m$$
$$1000 \; m = 1 \; k$$

From $\quad$ unit ratio $= \dfrac{\textit{New unit}}{\textit{original unit}}$

$$= \frac{1\,km}{1000\,m}$$

$$3000\,m = \cancel{3000\,m} \times \frac{1\,km}{\cancel{1000\,m}}$$

$$\therefore \quad 3000\,m = 3\,km$$

OR   3000 meters to kilometers

$$1\,km = 1000\,m$$

$$1000\,m = 1\,km$$

Divide by 1000 on both sides

$$\frac{\cancel{1000}\,m}{\cancel{1000}} = \frac{1\,km}{1000}$$

$$1\,m = \frac{1}{1000}km$$

Multiply by 3000 on both sides

$$3000m = \frac{1}{\cancel{1000}}\,km \times \cancel{3000}$$

$$3000m = 3\,km$$

U.S units of weight

1 ton = 2000 pounds.   In unit ratio $= \dfrac{1\,ton}{2000\,pounds}$

Or $\dfrac{2000\,pounds}{1\,ton}$

1 pound = 16 ounces.   In unit ratio $= \dfrac{1\,pound}{16\,ounces}$

Or $\dfrac{16\,ounces}{1\,pound}$

Examples 14.2

Convert 5 pounds into ounces

Solution

From   1 pound = 16 ounces

$$\text{Unit ratio} = \frac{New\ unit}{original\ unit}$$

$$\text{Unit ratio} = \frac{16\ ounces}{1\ pound}$$

$$5\ \text{pounds} = \frac{16\ ounces}{1\ \cancel{pound}} \times 5\ \cancel{pounds}$$

5 pounds = 80 ounces

OR   if 1 pound = 16 ounces

5 pounds = (16 × 5) ounces

5 pounds = 80 ounces

U.S units of length

1 mile = 5280 feet.   In unit ratio $= \dfrac{1\ mile}{5280\ feet}$

Or $\dfrac{5280\ feet}{1\ mile}$

1 mile = 1760 yards.   In unit ratio $= \dfrac{1\ mile}{1760\ yards}$

Or $\dfrac{1760\ yards}{1\ mile}$

1 yard = 36 inches.   In unit ratio $= \dfrac{1\ yard}{36\ inches}$

Or $\dfrac{36\ inches}{1\ yard}$

1 yard = 3 feet   In unit ratio $= \dfrac{1\ yard}{3\ feet}$

$$Or \quad \frac{3\,feet}{1\,yard}$$

1 foot = 12 inches.   In unit ratio $= \dfrac{1\,foot}{12\,inches}$

$$Or \quad \frac{12\,inches}{1\,foot}$$

Examples 14.3
(i) Convert 3 feet into inches
Solution

From 1 foot = 12 inches

$$Unit\ ratio = \frac{12\,inches}{1\,foot}$$

$$3\ feet = \frac{12\,inches}{1\,\cancel{foot}} \times 3\ \cancel{feet}$$
3 feet = 36 inches

Or   if 1 foot = 12 inches
3 feet = (12 × 3) inches
3 feet = 36 inches

(ii) Convert 90 feet into yards
Solution

From 1 yard = 3 feet

$$Unit\ ratio = \frac{1\,yard}{3\,feet}$$

$$90\ feet = \frac{1\,yard}{3\,\cancel{feet}} \times \cancel{90\ feet}^{30}$$
90 feet = 30 yards

Or   From 1 yard = 3 feet
3 feet = 1 yard

Divide by 3 on both sides

$$\frac{3\,feet}{3} = \frac{1\,yard}{3}$$

$$1 \text{ foot} = \frac{1}{3}\,yard$$

$$90 \text{ feet} = \left(90 \times \frac{1}{3}\right) \text{ yards}$$

$$90 \text{ feet} = 30 \text{ yards}$$

U.S units of capacity

1 gallon = 4 quarts. In unit ratio $= \dfrac{1\,gallon}{4\,quarts}$

Or $\dfrac{4\,quarts}{1\,gallon}$

1 quart = 2 pints. In unit ratio $= \dfrac{1\,quart}{2\,pints}$

Or $\dfrac{2\,pints}{1\,quart}$

1 pint = 2 cups in unit ratio $= \dfrac{1\,pint}{2\,cups}$

Or $\dfrac{2\,cups}{1\,pint}$

1 cup = 8 fluid ounces in unit ratio $= \dfrac{1\,cup}{8\,fluid\,ounces}$

Or $\dfrac{8\,fluid\,ounces}{1\,cup}$

1 table spoon = 3 tea spoons. In unit ratio $= \dfrac{1\,table\,spoon}{3\,tea\,spoons}$

Or $\dfrac{3\,tea\,spoons}{1\,table\,spoon}$

Examples 14.4

Convert *80* fluid ounces to pints

Solution

$$\text{Unit ratio} = \frac{New\ unit}{original\ unit}$$

From   1 cup = 8 fluid ounces

$$\text{Unit ratio} = \frac{1\ cup}{8\ fluid\ ounces} \qquad (1)$$

From   1 pint = 2 cups

$$\text{Unit ratio} = \frac{1\ pint}{2\ cups} \qquad (2)$$

Using (1) and (2)

$$80\ \text{fluid ounces} = \frac{1\ cup}{8\ fluid\ ounces} \times \frac{1\ pint}{2\ cups} \times 80\ \text{fluid ounces}$$

80 fluid ounces = 5 pints

Or   80 fluid ounces to pints

From   1 cup = 8 fluid ounces

8 fluid ounces = 1 cup

Divide by *8* on both sides

$$1\ \text{fluid ounce} = \frac{1}{8}\ \text{cup}$$

$$80\ \text{fluid ounces} = \left(80 \times \frac{1}{8}\right)\ \text{cups}$$

80 fluid ounces = 10 cups

Also from   1 pint = 2 cups

2 cups = 1 pint

Divide by *2* on both sides

$$1\ \text{cup} = \frac{1}{2}\ \text{pint}$$

$$10\ \text{cups} = \left(10 \times \frac{1}{2}\right)\ \text{pints}$$

10 cups = 5 pints

80 fluid ounces = 10 cups

∴ 80 fluid ounces = 10 cups = 5 pints

U.S units of health care units

1 milliliter = 15 drops   in unit ratio = $\dfrac{1\ milliliter}{15\ drops}$

Or $\dfrac{15\ drops}{1\ milliliter}$

1 tea spoon = 60 drops   in unit ratio = $\dfrac{1\ tea\ spoon}{60\ drops}$

Or $\dfrac{60\ drops}{1\ tea\ spoon}$

1 milligram = 1000 micrograms.

In unit ratio $\dfrac{1\ milligram}{1000\ micrograms}$ or $\dfrac{1000\ micrograms}{1\ milligram}$

Examples 14.5

Convert 30 drops to tea spoons

Solution

From   1 tea spoon = 60 drops

Unit ratio = $\dfrac{1\ tea\ spoon}{60\ drops}$

$30\ drops = \dfrac{1\ tea\ spoon}{60\ drops} \times 30\ drops$

$30\ drops = \dfrac{1}{2}\ tea\ spoon$

$30\ drops = 0.5\ tea\ spoon$

Units of time

1 minute = 60 seconds

1 hour = 60 minutes

1 day = 24 hours

1 week = 7 days

1 quarter = 3 months

1 year = 365 days (366 days leap year)

$$= 52 \text{ weeks} = 12 \text{ months}$$
$$1 \text{ decade} = 10 \text{ years}$$
$$1 \text{ century} = 100 \text{ years} = 10 \text{ decades}$$
$$1 \text{ millennium} = 1000 \text{ years} = 10 \text{ centuries}$$

Example 14.6

Convert *1* week into hours

Solution

$$1 \text{ week} = 7 \text{ days}$$
$$1 \text{ day} = 24 \text{ hours}$$
$$7 \text{ days} = 7 \times 24 \text{ hours}$$
$$1 \text{ week} = 168 \text{ hours}$$

14-2 Measuring temperature in Fahrenheit *F* and Celsius *C*

Fahrenheit $\quad F = \dfrac{9}{5} c + 32$

Celsius $\quad C = \dfrac{5}{9} (F - 32)$

Example 14.7

Convert 50°F into °C

Solution

$$\text{From} \quad C = \frac{5}{9} (F - 32)$$

$$C = \frac{5}{9} (50 - 32)$$

$$= \frac{5}{9} \times 18$$
$$= 10°C$$

14-3 Total cost (c) and simple interest (i)

Total cost (c) is a product of number of units (*n*) and cost per unit (r) $c = nr$.

Simple interest (i) is the product of principal (p), rate (r) and time (t). $i = prt$

Examples 14.8

(a) A student buys 10 text books each at the cost of $60. Find the total cost for all text books.

Solution

$$\text{Total cost } c = nr, \quad \text{where } n = 10 \text{ and } r = 60$$
$$= 10 \times 60$$
$$= \$600$$

(b) Find the simple interest on $500 at 17.5% per year for 4 years.

Solution

$$\text{From} \quad i = prt$$
$$= 500 \times \frac{17.5}{100} \times 4$$
$$= \$350$$

(c) Rose purchases a new car for a total price of $10,000. She gets a 10% discount plus $1200 trade in for her old car. She also gets a loan with 15% annual interest rate for 40 months to pay off the balance.

What are Rose's monthly payments to pay off the loan? Give your answer to the nearest cent.

Solution

$$10\% \text{ discount plus } \$1200 \text{ trade in}$$
$$= \frac{10}{100} \times 10000 + \$1200$$
$$= \$1000 + \$1200$$
$$= \$2200$$
$$\text{Balance} = \$10000 - \$2200$$
$$= \$7800$$

Interest on loan $(i) = prt$

$$= 7800 \times \frac{15}{100} \times \frac{40}{12}$$

$$= \$3900$$

Total to be paid = principal + interest

$$= \$7800 + \$3900$$

$$= \$11700$$

∴ Monthly payments

$$= \frac{11700}{40}$$

$$= \$292.50$$

Exercise 14A

(1) Convert *200* meters into kilometers

(2) Convert *200* centimeters to meters

(3) Convert *20* pounds into ounces

(4) Convert *500* pounds into tones

(5) Convert *50* feet into inches

(6) Convert *150* feet into yards

(7) Convert *12* inches into yards

(8) Convert *640* fluid ounces to pints

(9) Convert *5* gallons into cups

(10) Convert *6* drops to tea spoons

(11) Convert *600* micrograms into milligram

(12) Convert *2* week into hours

(13) How many weeks are in one century?

(14) Convert *122*°F into °C

(15) Convert 25°C into °F

(16) A student buys 15 text books each at the cost of $120. Find the total cost for the text books.

(17) A student buys 10 dozens of text books each book at the cost of $30. Find the total cost for all text books.

(18) Find the simple interest on $777 at $9\frac{3}{4}$% per year for 4 years.

(19) James borrows $1500 at 20% for 73 days. Find the interest due and the total amount paid for 73 days.

(20) A country's tax rate is 10% and the state's tax rate is 20%. 2 companies have goods worth $1200 each. Find the total amount sold for all the goods in 2 companies.

(21) Jane purchases a new car for a total price of $30,000. She gets a 15.5% discount plus $2000 trade in for her old car. She also gets a loan with 20% annual interest rate for 60 months to pay off the balance.
What are Jane's monthly payments to pay off the loan? Give your answer to the nearest cent.

(22) Peter and John began driving along a highway at the same time, same place and in the same direction. Peter drove at an average speed of 50mph and John drove at an average speed of 40mph. How far apart were they after 2.5 hours?

(23) Alex drove north along a highway at an average speed of 60mph. Josh drove south from the same place at the same time as Alex and at an average speed of 40mph. How far apart were they after 0.75 hour?

(24) Dona took 6 hours to do a page of math home work exercises. Lona took 2 hours to do the same number of exercises. If Dona's average rate was q exercises per hour. What was Lona's average rate in terms of q? Given, work (A) = rate (r) × time (t).

SOLUTIONS FOR EXERCISE 14A

(1)   *200* meters to kilometers

$$1 \ km = 1000 \ m$$

From   unit ratio $= \dfrac{New \ unit}{original \ unit}$

$$= \dfrac{1 \ km}{1000 \ m}$$

$$200m = 200 \ m \times \dfrac{1 \ km}{1000 \ m}$$

$$= \dfrac{2 \ km}{10}$$

$$\therefore 200 \ m = 0.2 \ km$$

OR   *200* meters to kilometers

$$1 \ km = 1000 \ m$$

$$1000 \ m = 1 \ km$$

Divide by *1000* on both sides

$$\dfrac{1000m}{1000} = \dfrac{1 \ km}{1000}$$

$$1m = \dfrac{1}{1000} \ km$$

Multiply by *200* on both sides

$$200m = \dfrac{1}{1000} \ km \times 200$$

$$= \dfrac{2}{10} \ km$$

$$\therefore 200m = 0.2 \ km$$

(2)   *200* centimeters to meters

$$1 \ m = 100 \ cm$$

Divide by *100* on both sides

$$1 \; cm = \frac{1}{100} \; m$$

$$200 \; cm = \frac{1}{100} \times 200 \; m$$
$$\therefore 200 \; cm = 2 \; m$$

(3) *20* pounds into ounces

From 1 pound = 16 ounces

$$\text{Unit ratio} = \frac{New \; unit}{original \; unit}$$

$$\text{Unit ratio} = \frac{16 \; ounces}{1 \; pound}$$

$$20 \text{ pounds} = \frac{16 \; ounces}{1 \; \cancel{pound}} \times 20 \; \cancel{pounds}$$
$$20 \text{ pounds} = 320 \text{ ounces}$$

OR if 1 pound = 16 ounces

20 pounds = (16 × 20) ounces

20 pounds = 320 ounces

(4) *500* pounds into tons

1 ton = 2000 pounds

2000 pounds = 1 ton

Divide by *2000* on both sides

$$1 \text{ pound} = \frac{1}{2000} \text{ ton}$$

$$500 \text{ pounds} = 500 \times \frac{1}{2000} \text{ ton}$$
$$= \frac{1}{4} \text{ ton}$$
$$500 \text{ pounds} = 0.25 \text{ ton}$$

(5) *50* feet into inches

From 1 foot = 12 inches

$$\text{Unit ratio} = \frac{12\ inches}{1\ foot}$$

$$50\ \text{feet} = \frac{12\ inches}{1\ \cancel{foot}} \times 50\ \cancel{\text{feet}}$$

50 feet = 600 inches

Or   if *1* foot = *12* inches

50 feet = (12 × 50) inches

50 feet = 600 inches

(6)   *150* feet into yards

From 1 yard = 3 feet

$$\text{Unit ratio} = \frac{1\ yard}{3\ feet}$$

$$150\ \text{feet} = \frac{1\ yard}{3\ \cancel{feet}} \times \cancel{150\ feet}^{\,50}$$

150 feet = 50 yards

Or   From 1 yard = 3 feet

3 feet = 1 yard

Divide by *3* on both sides

$$\frac{3\ feet}{3} = \frac{1\ yard}{3}$$

$$1\ \text{foot} = \frac{1}{3}\ \text{yard}$$

$$150\ \text{feet} = \left(150 \times \frac{1}{3}\right)\ \text{yards}$$

150 feet = 50 yards

(7)   *12* inches into yards

1 yard = 36 inches

36 inches = 1 yard

Divide by *36* on both sides

$$1 \text{ inch} = \frac{1}{36} \text{ yard}$$

$$12 \text{ inches} = 12 \times \frac{1}{36} \text{ yard}$$

$$= \frac{1}{3} \text{ yard}$$

$$\therefore 12 \text{ inches} = 0.33 \text{ yard}$$

(8)　*640* fluid ounces to pints

$$\text{Unit ratio} = \frac{New\ unit}{original\ unit}$$

From　1 cup = 8 fluid ounces

$$\text{Unit ratio} = \frac{1\ cup}{8\ fluid\ ounces} \qquad (1)$$

From　1 pint = 2 cups

$$\text{Unit ratio} = \frac{1\ pint}{2\ cups} \qquad (2)$$

Using (1) and (2)

$$640 \text{ fluid ounces} = \frac{1\ cup}{8\ fluid\ ounces} \times \frac{1\ pint}{2\ cups} \times 640 \text{ fluid ounces}$$

640 fluid ounces = 40 pints

OR　*640* fluid ounces to pints
From　1 cup = 8 fluid ounces
8 fluid ounces = 1 cup
Divide by 8 on both sides

$$1 \text{ fluid ounce} = \frac{1}{8} \text{ cup}$$

$$640 \text{ fluid ounces} = \left(640 \times \frac{1}{8}\right) \text{ cups}$$

640 fluid ounces = 80 cups
Also from　1 pint = 2 cups

2 cups = 1 pint

Divide by 2 on both sides

$$1 \text{ cup} = \frac{1}{2} \text{ pint}$$

$$80 \text{ cups} = \left(80 \times \frac{1}{2}\right) \text{ pints}$$

80 cups = 40 pints

∴ 640 fluid ounces = 80 cups = 40 pints

(9) *5* gallons into cups

From   1 gallon = 4 quarts

5 gallon = (4 × 5) quarts

5 gallon = 20 quarts

From   1 quart = 2 pints

20 quarts = (2 × 20) pints

20 quarts = 40 pints

From   1 pint = 2 cups

40 pints = (2 × 40) cups

40 pints = 80 cups

∴ 5 gallon = 20 quarts = 40 pints = 80 cups

(10) *6* drops to tea spoons

From 1 tea spoon = 60 drops

$$\text{Unit ratio} = \frac{1 \text{ tea spoon}}{60 \text{ drops}}$$

$$6 \text{ drops} = \frac{1 \text{ te spoon}}{60 \text{ drops}} \times 6 \text{ drops}$$

$$6 \text{ drops} = \frac{1}{10} \text{ teaspoon}$$

6 drops = 0.1 teaspoon

(11) *600* micrograms into milligram

From 1 milligram = 1000 micrograms

$$1 \text{ microgram} = \frac{1}{1000} \text{ milligram}$$

$$600 \text{ micrograms} = \cancel{600} \times \frac{1}{\cancel{1000}} \text{ milligrams}$$

$$600 \text{ micrograms} = \frac{6}{10} \text{ milligram}$$

$$600 \text{ micrograms} = 0.6 \text{ milligram}$$

(12)   2 week into hours

$$1 \text{ week} = 7 \text{ days}$$
$$2 \text{ week} = 14 \text{ days}$$
$$1 \text{ day} = 24 \text{ hours}$$
$$14 \text{ days} = 14 \times 24 \text{ hours}$$
$$\therefore \quad 2 \text{ week} = 336 \text{ hours}$$

(13)   From   1 year = 52 weeks

$$100 \text{ years} = (52 \times 100) \text{ weeks}$$
$$100 \text{ years} = 5200 \text{ weeks}$$

From   1 century = 100 years

$$\therefore \quad 1 \text{ century} = 100 \text{ years} = 5200 \text{ weeks}$$

(14)   122°F into °C

$$\text{From} \quad C = \frac{5}{9}(F - 32)$$

$$C = \frac{5}{9}(122 - 32)$$

$$= \frac{5}{9} \times 90$$
$$= 50°C$$

(15)   25°C into °F

$$\text{From} \quad F = \frac{9}{5}c + 32$$

$$= \frac{9}{5} \times 25 + 32$$
$$= 45 + 32$$
$$= 77°F$$

(16)　Total cost $c = nr$, where $n = 15$ and $r = 120$
$$= 15 \times 120$$
$$= \$1800$$

(17)　Total cost $c = nr$, where $n = 10$ dozens and $r = 30$
$$= 10 \times 12 \times 30$$
$$= \$3600$$

(18)　From　$i = prt$
$$= 777 \times 9\frac{3}{4}\% \times 4$$
$$= 777 \times \frac{39}{4}\% \times 4$$
$$= 777 \times \frac{39}{400} \times 4$$
$$= \$303.03$$

(19)　Interest due $= \$1500 \times \frac{20}{100} \times \frac{73}{365}$
$$= \$60$$
Total amount paid for 73 days
$$= \$1500 + \$60$$
$$= \$1560$$

(20)　Total taxes on goods
$$= (10\% + 20\%) \times 1200 \times 2$$
$$= \frac{30}{100} \times 1200 \times 2$$

$$= \$720$$

Total amount sold

$$= (\$1200 \times 2) + \$720$$
$$= \$2400 + \$720$$
$$= \$3120$$

(21) *15.5% discount plus $2000 trade in*

$$= \frac{15.5}{100} \times 30000 + \$2000$$
$$= \$4650 + \$2000$$
$$= \$6650$$

Balance $= \$30000 - \$6650$
$$= \$23350$$

Interest on loan $(i) = prt$

$$= 23350 \times \frac{20}{100} \times \frac{60}{12}$$
$$= \$23350$$

Total to be paid $=$ principal $+$ interest
$$= \$23350 + \$23350$$
$$= \$46700$$

$\therefore$ Monthly payments $= \dfrac{46700}{60}$

$$= 778.333333$$
$$= \$778.33$$

(22) For Peter

Distance (d) = speed (s) $\times$ time (t)

$$= 50 \times 2.5$$
$$= 125 \; mi$$

For John $\qquad d = st$

$$= 40 \times 2.5$$
$$= 100 \; mi$$

Distance apart $= 125 - 100$

$$= 25 \; mi$$

(23) For Alex

$$\text{Distance } (d) = \text{speed } (s) \times \text{time } (t)$$
$$= 60 \times 0.75$$
$$= 45 \; mi$$

For Josh $d = st$
$$= 40 \times 0.75$$
$$= 30 \; mi$$

Distance apart $= 45 + 30$
$$= 75 \; mi$$

(24) For Dona

Work(A) = rate (r) × time (t)
$$= q \times 6$$
$$= 6q$$

For Lona $A = rt$
$$= r \times 2$$
$$= 2r$$

Since same work done
$$2r = 6q$$
$$r = \frac{6q}{2}$$
$$r = 3q$$

Exercise 14B

(1) Convert 50 meters into kilometers

(2) Convert 7000 centimeters to meters

(3) Convert 32 pounds into ounces

(4) Convert 800 pounds into tons

(5) Convert 120 feet into inches

(6) Convert 720 feet into yards

(7) Convert 20 inches into yards

(8) Convert 800 fluid ounces to pints

(9) Convert 12 gallons into cups

(10) Convert 240 drops to tea spoons

(11) Convert 10000 micrograms into grams

(12) Convert 5 week into hours

(13) How many weeks are in two centuries?

(14) Convert 212°F into °C

(15) Convert 50°C into °F

(16) A student buys 50 text books each at the cost of $110. Find the total cost for the text books.

(17) A student buys 12 dozens of text books each book at the cost of $50. Find the total cost for all text books.

(18) Find the simple interest on $850 at $2\frac{3}{5}\%$ per year for 2 years.

(19) Daniel borrows $2200 at 30% for 146 days. Find the interest due and the total amount paid for 146 days.

(20) A country's tax rate is 10% and the state's tax rate is 20%. 3 companies have goods worth $2400 each. Find the total amount sold for all the goods in 3 companies.

(21) Ron purchases a new car for a total price of $50,000. He gets a 18.2% discount plus $2500 trade in for his old car. He also gets a loan with 25% annual interest rate for 36 months to pay off the balance.

315

What are Ron's monthly payments to pay off the loan? Give your answer to the nearest cent.

(22) Eron and Patrick began driving along a highway at the same time, same place and in the same direction. Eron drove at an average speed of *85mph* and Patrick drove at an average speed of *65mph*. How far apart were they after *4.5* hours?

(23) Eric drove east along a highway at an average speed of *75mph*. Mike drove west from the same place at the same time as Eric and at an average speed of *55mph*. How far apart were they after *1.5* hours?

(24) Ruth took *12* hours to do a page of science home work exercises. Sam took *9* hours to do the same number of exercises. If Ruth's average rate was *y* exercises per hour. What was Sam's average rate in terms of *y*? Given, work (A) = rate (r) × time (t).

CHAPTER 15

GEOMETRY

15-1 Sides and angles of a polygon
Angles
<u>Vertical angles</u>

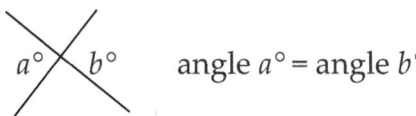

angle $a° =$ angle $b°$

<u>Adjacent angles</u> are two angles with common vertex and common sides and no common interior

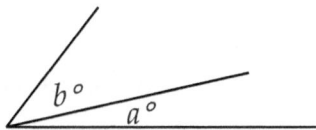

<u>Right angle</u> is an angle whose measure is $90°$

<u>An acute angle</u> is an angle whose measure is less than $90°$ $(0<\theta<90°)$

<u>Obtuse angle</u> is an angle whose measure is more than $90°$ and less than $180°$ $(90°<\theta<180°)$

<u>A straight angle</u> is an angle whose measure is $180°$

 Also the sum of all angles on a straight line is 180

Angles $a° + b° + c° = 180$

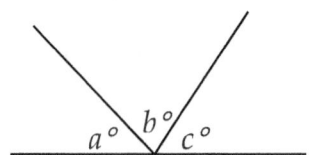

A <u>reflex angle</u> is an angle whose measure is more than *180°* and less than *360° (180°<θ<360°)*

<u>Complimentary angles</u> are two angles whose measure totals *90°*

<u>Supplementary angles</u> are two angles whose measure totals *180°*

<u>Congruent angles</u> are angles of equal measure

A <u>transversal line</u> is a line that crosses two parallel lines forming eight angles.

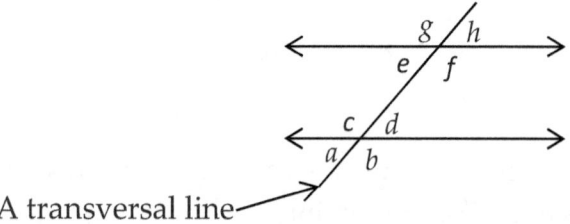

A transversal line

<u>Alternate interior angles</u> are like angles *d* and *e* (*d°* = *e°*)

<u>Alternate exterior angles</u> are like angles *b* and *g* (*b°* = *g°*)

A Polygon is a figure with the same number of sides as angles

Table 15.1

Examples of polygons	sides / interior angles
Equilateral triangles	3 / 60°
Quadrilateral	4 / 90°
Regular Pentagon	5 / 108°
Regular Hexagon	6 / 120°
Regular Heptagon	7 / 128.57°
Regular Octagon	8 / 135°
Regular Nonagon	9 / 140°
Regular Decagon	10 / 144°
Regular Hectagon	100 / 176.4°
Regular Megagon	10^6 / 179.9996°
Regular Googolgon	10^{100} / 180°

An <u>equilateral polygon</u> is a polygon with all its sides equal.
An <u>equiangular polygon</u> is a polygon whose angles are equal.
<u>Regular polygon</u> is a polygon that is equilateral and also equiangular.
Each <u>exterior angle</u> of a regular polygon of n sides measures

$$\frac{360°}{n} \text{ degrees}$$

<u>Perimeter</u> (p) is a measure around the figure or an object the perimeter of a circle is called <u>circumference</u> (c).
<u>Area</u> (A) is a measure of the interior of the figure or an object.
The area of a regular polygon equals one half the product of length of the apothem (a) and the perimeter (p)

$$\textbf{Area} = \frac{1}{2}ap$$

r is a radius

Examples 15.1

a(i)

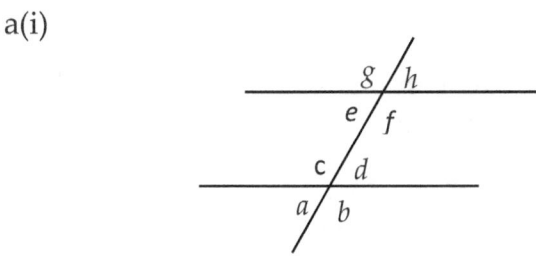

Given $a° = 60°$; find all other angles and their relationship between each other.

$a° = d° = 60°$ Vertical angles
$b° = 180° - a°$ Straight angles(angles on line = $180°$)
$b° = 180° - 60°$
$b° = 120°$
$b° = c° = 120°$ Vertical angles
$d° = e° = 60°$ Alternate interior angles
$b° = g° = 120°$ Alternate exterior angles
$g° = f° = 120°$ Vertical angles
$h° = 180° - f°$ Straight angles(angles on line = $180°$)
$h° = 180° - 120°$
$h° = 60°$

(ii) Which angles are congruent angles.
    $a°, d°, e°$ and $h°$ are congruent angles also
    $b°, c°, f°$ and $g°$ are congruent angles

b(i) Exterior angle of a regular pentagon
$$= \frac{360°}{5}$$
$$= 72°$$

(ii) Exterior angle of a regular hexagon
$$= \frac{360°}{6}$$
$$= 60°$$

(iii) Exterior angle of a regular octagon
$$= \frac{360°}{8}$$
$$= 45°$$

c(i) The length of the hexagon is *10m* and its apothem is *7m*
    Find the area of the hexagon.

Solution

$$\text{Area} = \frac{1}{2}ap$$

But perimeter $p = 10m \times 6$ sides

$$= 60m$$

$$\text{Area} = \frac{1}{2} \times 7m \times 60m$$

$$= 7m \times 30m$$

$$= 210m^2$$

(ii) The side of a regular pentagon is $6m$ and the radius is $5m$. find the area of a regular pentagon.

Solution

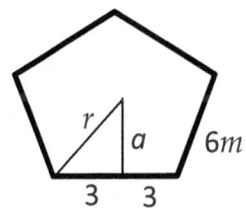

From Pythagorean theory

$$r^2 = a^2 + 3^2$$

$$5^2 = a^2 + 3^2$$

$$25 = a^2 + 9$$

$$25 - 9 = a^2$$

$$16 = a^2$$

$$4^2 = a^2$$

$$4 = a$$

$\therefore$ apothem $(a) = 4m$

Perimeter $P = 6m + 6m + 6m + 6m + 6m$

$$= 30m$$

From Area $= \frac{1}{2}ap$

$$= \frac{1}{2} \times 4m \times 30m$$

$$\text{Area} = 60m^2$$

15-2 Shapes of figures or objects

<u>Triangles</u> are closed 3 sided geometric figure

Area of a triangle $= \frac{1}{2}bh$
where b is the base and h is the height.

Its perimeter is $\quad a + b + c$

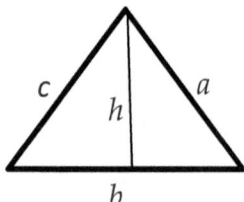

The sum of all interior angles of a triangle is *180°*
<u>Scalene triangle</u> is a triangle with no equal sides
<u>Isosceles triangle</u> is a triangle with 2 equal sides
<u>Equilateral triangle</u> is a triangle with 3 equal sides
<u>Obtuse triangle</u> is a triangle with one obtuse angle $\theta > 90°$
<u>Acute triangle</u> is a triangle with 3 acute angles $\theta < 90°$
<u>Right triangle</u> is a triangle with right angle *(90°)*

2 triangles are said to be <u>congruent</u> if it has the following:

(i) SAS (side angle side)

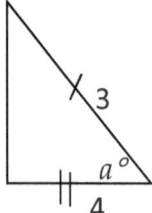 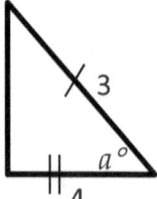

(ii) ASA (angle side angle)

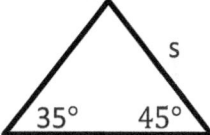 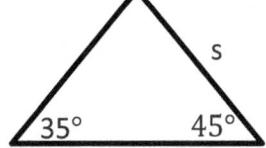

(iii) SSS (side side side)

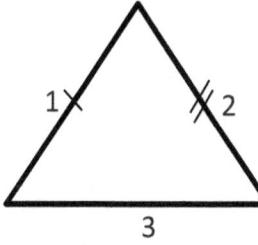 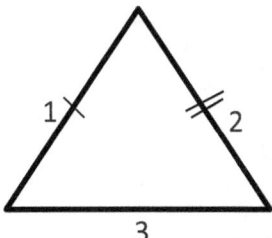

A rectangle

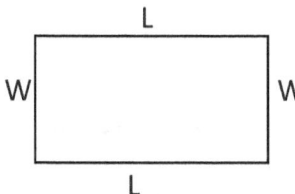

Area $A = LW$

Perimeter $P = 2L + 2W$

A square

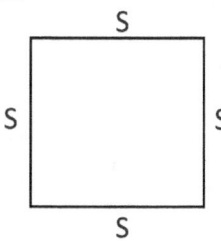

$$A = S \times S = S^2$$
$$P = S + S + S + S = 4S$$

Parallelogram

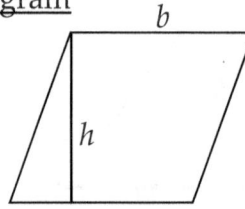

$$A = bh$$

Trapezoid

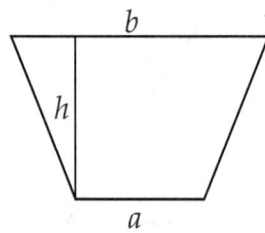

$$A = \frac{1}{2}(a + b)h$$

Circle

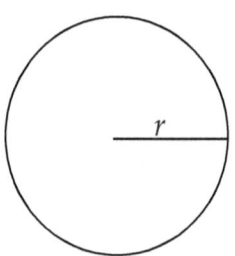

r is the radius

$$A = \pi r^2$$

Circumference $C = \pi D = 2\pi r$

where D is diameter, and $D = 2r$

Given the radius r, and the central angle n, the arc length AB is given by $\frac{n}{360°}2\pi r$.

$(\pi = 3.14)$

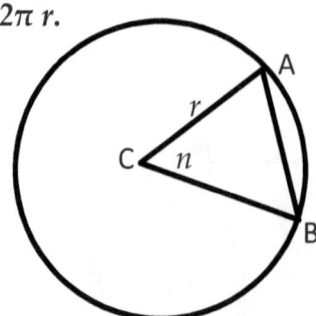

Rectangular solid

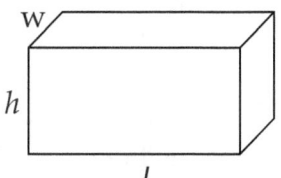

Volume $V = Lwh$

Cube

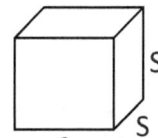

$V = S \times S \times S = S^3$

Cylinder

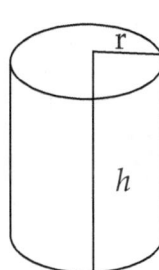

Volume $V = \pi r^2 h$

Cone

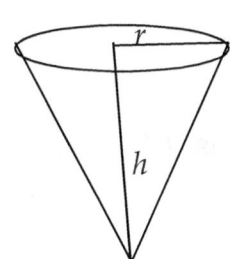

Volume $V = \dfrac{1}{3} \pi r^2 h$

Pyramid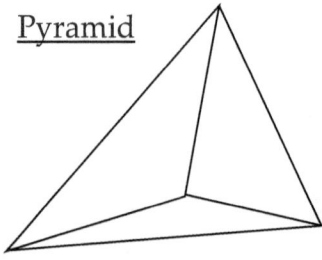

Volume $V = \dfrac{1}{3}Bh$

Where B is the area of the base,
h is the height

Sphere

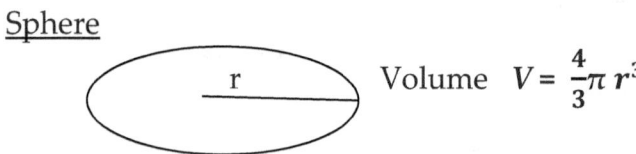

Volume $V = \dfrac{4}{3}\pi\, r^3$

Examples 15.2

a(i) Find the perimeter of an equilateral triangle with side
measure *7in*.

Solution

For equilateral triangle all sides are equal

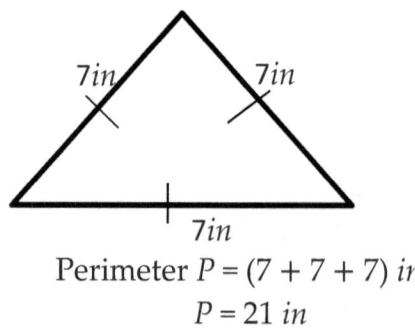

Perimeter $P = (7 + 7 + 7)\ in$
$P = 21\ in$

(ii) The vertex angle of an isosceles triangle is *70 °*
Find the base angle.

Solution

Let the base angle of an isosceles triangle be *x*

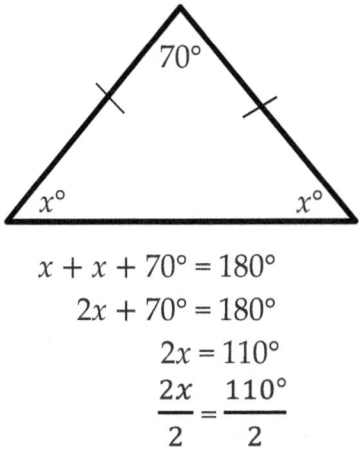

$$x + x + 70° = 180°$$
$$2x + 70° = 180°$$
$$2x = 110°$$
$$\frac{2x}{2} = \frac{110°}{2}$$

$$x = 55°$$

∴ the base angle is 55°

(iii) Find the perimeter and the area of a triangle bellow.

Solution

Let the hypotenuse of a triangle be c

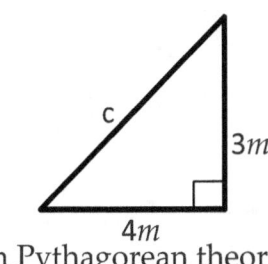

From Pythagorean theory
$$c^2 = 4^2 + 3^2$$
$$c^2 = 16 + 9$$
$$c^2 = 25$$
$$c^2 = 5^2$$
$$c = 5m$$

∴ the perimeter $= 5m + 4m + 3m$
$$= 12m$$

$$\text{Area} = \frac{1}{2}bh$$

$$= \frac{1}{2} \times 4m \times 3m$$
$$= 6m^2$$

(iv) The triangles area similar and the sides are proportion
Find a.

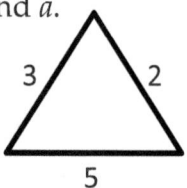 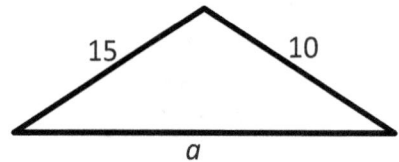

Solution

$$3 \times 5 = 15$$
$$2 \times 5 = 10$$
$$5 \times 5 = a$$
$$25 = a$$
$$\therefore \quad a = 25$$

(v) Two triangles said to be Congruent. Find b

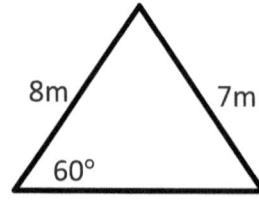 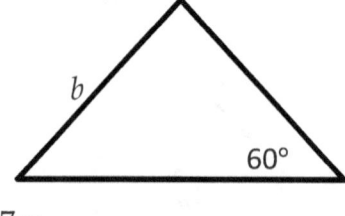

$$b = 7m$$

b(i) The perimeter of a rectangle is *28m*, its length is *8m* and its width is *2x*. find x

Solution

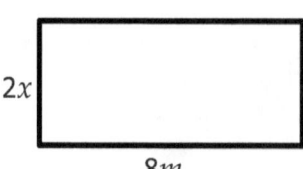

$$2x + 2x + 8 + 8 = 28$$
$$4x + 16 = 28$$
$$4x + 16 - 16 = 28 - 16$$
$$4x = 12$$

$$\frac{4x}{4} = \frac{12}{4}$$
$$x = 3m$$

(ii) The length of a rectangle is twice the width. Find the width given the area of a rectangle is $200m^2$.

Solution

Let the width of a rectangle be w

Length of a rectangle is $2w$

From the area $A = LW$

$$200 = 2w \times w$$
$$200 = 2w^2$$
$$\frac{200}{2} = \frac{2w^2}{2}$$

$$100 = w^2$$
$$w^2 = 10^2$$
$$w = 10$$

∴ the width of a rectangle is $10m$

(iii) The length of a rectangle is three more than the width. Find width given the perimeter of the rectangle is $10m$.

Solution

Let the width of a rectangle be x

The length of the rectangle is $x + 3$

From Perimeter $P = 2L + 2W$

$$10 = 2(x + 3) + 2x$$
$$10 = 2x + 6 + 2x$$
$$10 = 4x + 6$$

$$4 = 4x$$
$$x = 1$$

∴ the width of a rectangle is *1m*

(c) Find the area of the following figures:
   Given $a = 3m$, $b = 8m$, $h = 5m$ and diameter $D = 4m$

(i)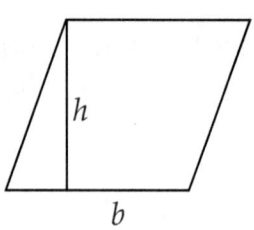

From area $A = bh$
$$A = 8m \times 5m$$
$$A = 40m^2$$

(ii)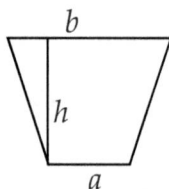

From area $A = \frac{1}{2}(a + b)h$

$$A = \frac{1}{2}(3 + 8)5$$

$$A = \frac{1}{2}(11)5$$

$$= \frac{55}{2}$$
$$A = 27.5m^2$$

(iii)

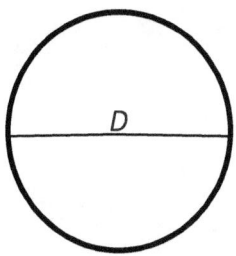

$$\text{Radius } r = \frac{D}{2}$$
$$= \frac{4}{2}$$
$$r = 2m$$
$$\text{From area } A = \pi r^2$$
$$A = \pi 2^2$$
$$A = 4\,\pi m^2$$
$$A = 12.56m^2$$

(d) Find the volume of the following: Given $w = 3m$, $L = 8m$, $h = 5m$ and for cylinder and cone diameter $D = 4m$.

(i)

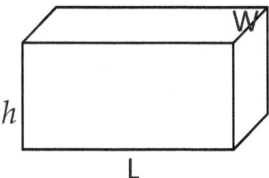

$$\text{Volume } V = Lwh$$
$$= 8 \times 3 \times 5$$
$$V = 120m^3$$

(ii)

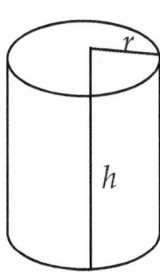

$$V = \pi r^2 h$$
$$\text{But } r = \frac{D}{2}$$

$$= \frac{4}{2}$$
$$r = 2m$$
$$V = \pi 2^2 * 5$$
$$V = 20\,\pi m^3$$
$$V = 62.8 m^3$$

(iii)

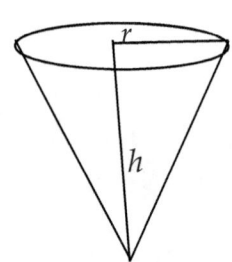

$$\text{Volume } V = \frac{1}{3}\pi r^2 h$$

$$\text{But } r = \frac{D}{2}$$

$$= \frac{4}{2}$$
$$r = 2m$$
$$V = \frac{1}{3}\pi 2^2 * 5$$

$$V = \frac{20\,\pi}{3}\, m^3$$
$$V = 20.\,933 m^3$$

(e) The following figures are filled with water.
   Given the diameter of the cylinder and cone $D = 10m$,
   $h = 9m$, $h_1 = 3m$, $h_2 = 6m$ and $r_2 = 4m$.   Find the

volume of the empty space that is not filled with water.

(i)

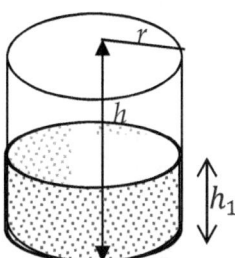

Volume of cylinder $V = \pi r^2 h$

$$\text{But } r = \frac{D}{2}$$

$$= \frac{10}{2}$$

$$r = 5m$$

$$V = \pi 5^2 * 9$$

$$V = 706.5m^3$$

Volume of water in the cylinder

$$V = \pi r^2 h_1$$

$$\text{But } r = \frac{D}{2}$$

$$= \frac{10}{2}$$

$$r = 5m$$

$$V = \pi 5^2 * 3$$

$$V = 235.5m^3$$

Volume of empty space in the cylinder

$$706.5m^3 - 235.5m^3 = 471m^3$$

Or   height of empty space $= h - h_1$

$$9m - 3m = 6m$$

$$V = \pi 5^2 * 6$$

$$V = 471m^3$$

(ii)

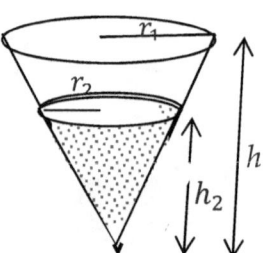

Volume of the cone $V = \dfrac{1}{3}\pi r_1{}^2 h$

But $r_1 = \dfrac{D}{2}$

$$= \dfrac{10}{2}$$

$r_1 = 5m$

$V = \dfrac{1}{3}\pi 5^2 * 9$

$V = 235.5 m^3$

Volume of water in the cone

$$V = \dfrac{1}{3}\pi\, r_2{}^2 h_2$$

$$V = \dfrac{1}{3}\pi 4^2 * 6$$

$$V = 100.48 m^3$$

Volume of empty space in the cone

$235.5 m^3 - 100.48 m^3 = 135.02 m^3$

(f) Find the arc length AB with radius $BC = 8m$

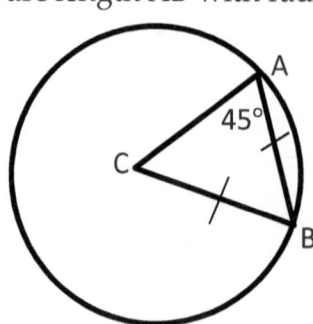

From arc length formula $= \dfrac{n}{360°}2\pi\, r$

$$= \dfrac{45°}{360°}2\pi*8m$$
$$= 2\pi m$$
$$= 6.28m$$

15-3 Equation of a circle, Parabola, Ellipse and Hyperbola

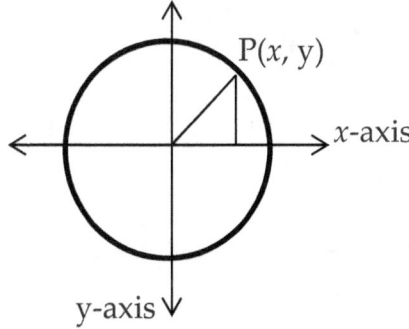

The equation of the **circle** with center at origin $(0,0)$
and radius r is $\quad x^2 + y^2 = r^2$

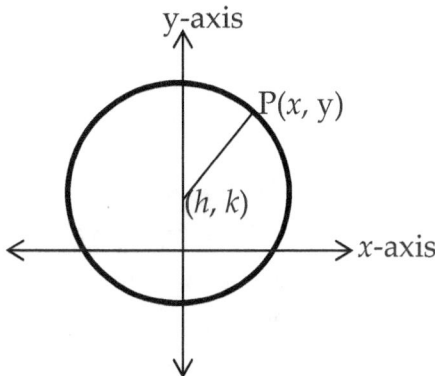

The equation of the circle with center at (h, k) and radius r
is the distance from any point on the circle (x, y) to the
center (h, k) is given by $(x - h)^2 + (y - k)^2 = r^2$
Also the equation of the circle can be given in this form

$$ax^2 + ay^2 + bx + cx + d = 0 \quad \text{where } a, b, c \text{ and } d \text{ are}$$
real number $a \neq 0$.

Examples 15.3
a(i) Find the equation of the circle with center at origin
and radius 7.
Solution

From $\quad x^2 + y^2 = r^2$
$$x^2 + y^2 = 7^2$$
$$x^2 + y^2 = 49$$

(ii) Find the center and radius of the equation of a circle
$x^2 + y^2 = 16$.
Solution

$$x^2 + y^2 = 16$$
$$x^2 + y^2 = 4^2$$
From $\quad x^2 + y^2 = r^2 \quad$ the center is at origin $(0, 0)$.
$$r^2 = 4^2$$
$\therefore$ Radius $r = 4$

(iii) Find equation of the circle with center at $(5, -10)$ and
with radius 3.
Solution

From $\quad (x - h)^2 + (y - k)^2 = r^2$
$$(x - 5)^2 + (y - ^-10)^2 = 3^2$$
$$(x - 5)^2 + (y + 10)^2 = 9$$
$$x^2 - 10x + 25 + y^2 + 20y + 100 = 9$$
$$x^2 + y^2 - 10x + 20y + 25 + 100 - 9 = 0$$
$$x^2 + y^2 - 10x + 20y + 116 = 0$$

(iv) Find the center and the radius of the circle with
equation $\quad 2x^2 + 2y^2 - 8x - 32y + 118 = 0$.
Solution

$$2x^2 + 2y^2 - 8x - 32y + 118 = 0$$

Divide each term by 2
$$x^2 + y^2 - 4x - 16y + 59 = 0$$
$$x^2 - 4x + y^2 - 16y + 59 = 0$$
Square half of the coefficient of x and y and add it on both sides.

$$x^2 - 4x + \left(\frac{-4}{2}\right)^2 + y^2 - 16y + \left(\frac{-16}{2}\right)^2 + 59 = \left(\frac{-4}{2}\right)^2 + \left(\frac{-16}{2}\right)^2$$
$$x^2 - 4x + (^-2)^2 + y^2 - 16y + (^-8)^2 + 59 = (^-2)^2 + (^-8)^2$$
$$x^2 - 4x + 4 + y^2 - 16y + 64 + 59 = 4 + 64$$
$$x^2 - 4x + 4 + y^2 - 16y + 64 = 4 + 64 - 59$$
$$(x - 2)^2 + (y - 8)^2 = 9$$

Compare with
$$(x - h)^2 + (y - k)^2 = r^2$$

The center $(h, k) = (2, 8)$

The radius $r^2 = 9$
$$= 3^2$$
$$r = 3$$

Parabola is a set of points formed with equal distance from a fixed point (the focus) and the fixed line (the directrix).
Vertex is a midpoint between focus (F) and directrix (D)
Axis is perpendicular to directric through the focus

$$PF = PD$$

Parabolas with vertices (vertexes) at origin $(0, 0)$.

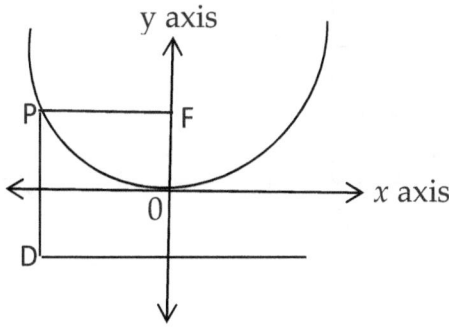

Standard forms of a parabola

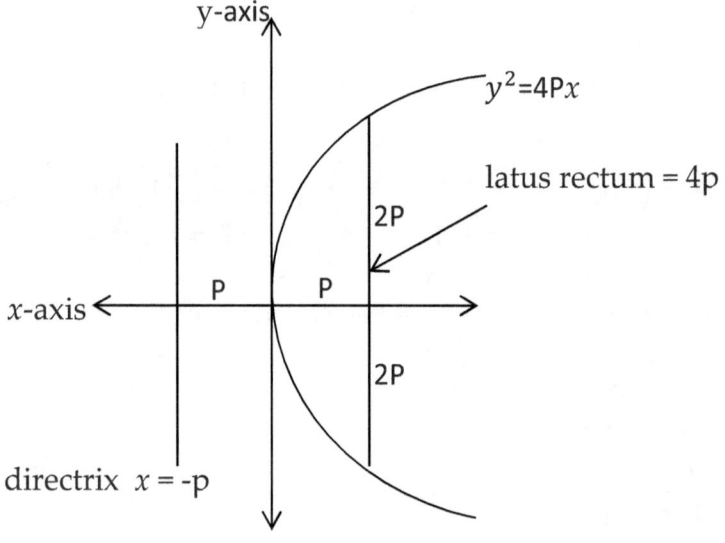

$$y^2 = 4px$$
vertex (0, 0) open right
directrix $x = -p$
focus (p, 0)
latus rectum = 4p

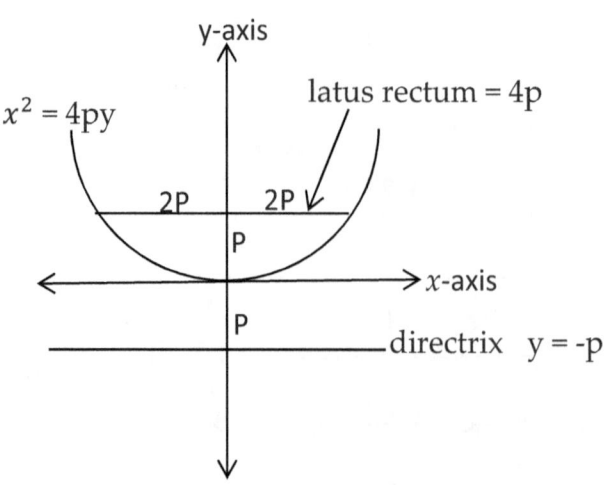

$x^2 = 4py$
vertex (0, 0) open upward
directrix y = -p
focus (0, p)
latus rectum = 4p

Parabolas with vertices (vertexes) at (h, k).

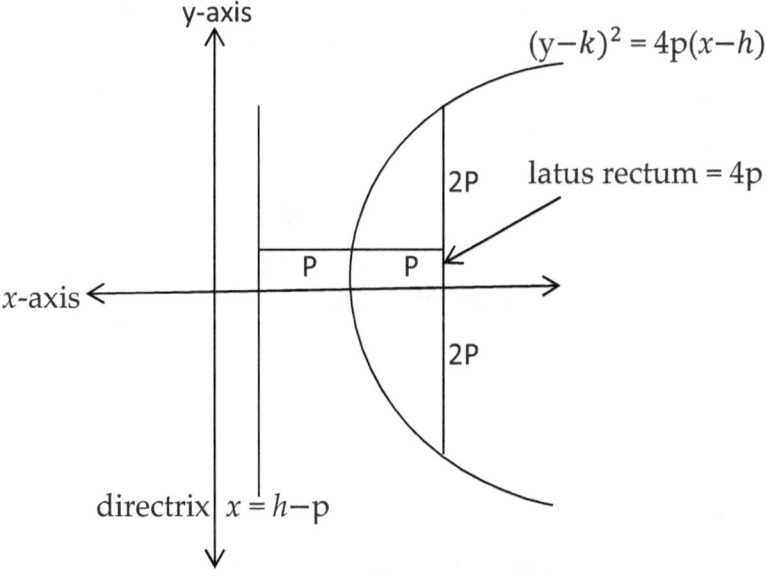

$(y-k)^2 = 4p(x-h)$
vertex (h, k) open to the right
directrix $x = h - p$
focus $(h + p, k)$
latus rectum = 4p

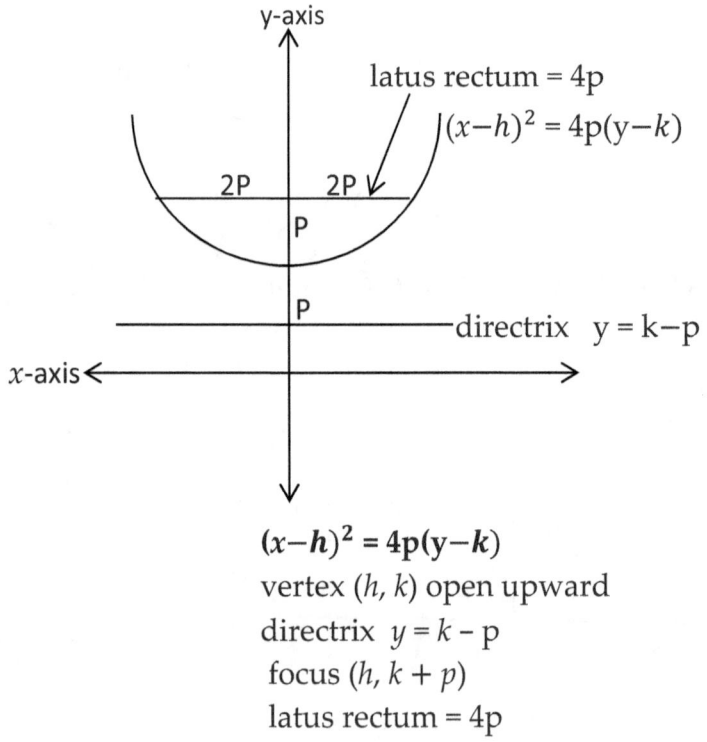

$(x-h)^2 = 4p(y-k)$
vertex (h, k) open upward
directrix $y = k - p$
focus $(h, k + p)$
latus rectum = 4p

Table 15.1

Table showing standard forms of a parabola.

$y^2 = 4px$	$(y-k)^2 = 4p(x-h)$
vertex $(0, 0)$ open right	vertex (h, k) open to the right
directrix $x = -p$	directrix $x = h - p$
focus $(p, 0)$	focus $(h + p, k)$
latus rectum = 4p	latus rectum = 4p
$x^2 = 4py$	$(x-h)^2 = 4p(y-k)$
vertex $(0, 0)$ open upward	vertex (h, k) open upward
directrix $y = -p$	directrix $y = k - p$
focus $(0, p)$	focus $(h, k + p)$
latus rectum = 4p	latus rectum = 4p

$y^2 = -4px$ vertex $(0, 0)$ open left directrix $x = p$ focus $(-p, 0)$ latus rectum $= 4p$	$(y-k)^2 = -4p(x-h)$ vertex (h, k) open left directrix $x = h + p$ focus $(h-p, k)$ latus rectum $= 4p$
$x^2 = -4py$ vertex $(0, 0)$ open downward directrix $y = p$ focus $(0, -p)$ latus rectum $= 4p$	$(x-h)^2 = -4p(y-k)$ vertex (h, k) open downward directrix $y = k + p$ focus $(h, k - p)$ latus rectum $= 4p$

Examples 15.3

b(i) Find the equation of a parabola with vertex at origin
   and focus $(3, 0)$.

Solution

   Let $P(x, y)$ be any point on a parabola
   Since the focus F is $(3, 0)$ and vertex $(0, 0)$,
   The directrix D is $x = ^-3$

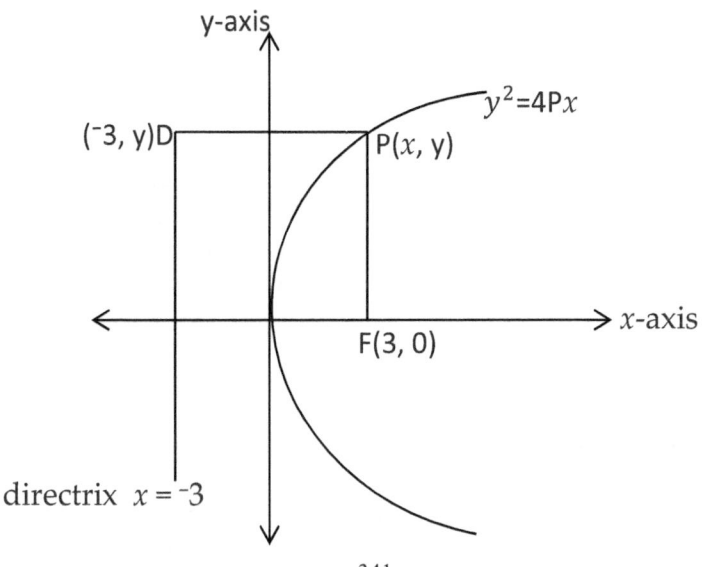

From $PF = PD$

$$\sqrt{(x-3)^2 + (y-0)^2} = \sqrt{(x-{}^-3)^2 + (y-y)^2}$$
$$(x-3)^2 + (y-0)^2 = (x+3)^2 + (y-y)^2$$
$$x^2 - 6x + 9 + y^2 = x^2 + 6x + 9 + 0^2$$
$$y^2 = 6x + 6x$$
$$y^2 = 12x$$

OR $Focus(p, 0) = (3, 0), \therefore p = 3$
From $y^2 = 4px$
$$y^2 = 4 * 3 * x$$
$$y^2 = 12x$$

(ii) Find the equation of a parabola with vertex at origin and focus $(0, 3)$.

Solution

Let $P(x, y)$ be any point on a parabola
Since the focus F is $(0, 3)$ and vertex $(0, 0)$,
The directrix D is $y = {}^-3$

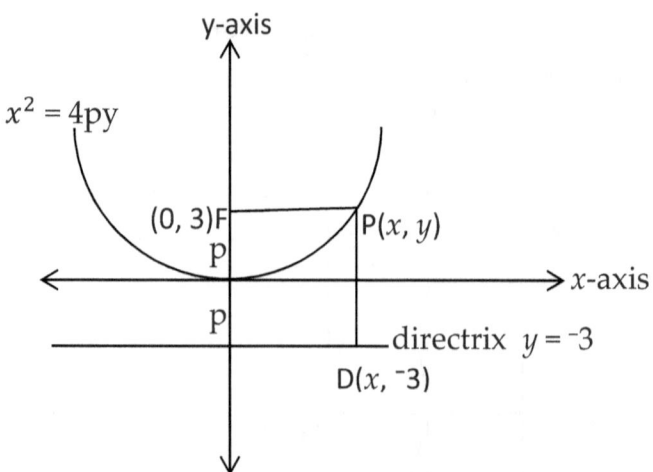

From $PF = PD$

$$\sqrt{(x-0)^2 + (y-3)^2} = \sqrt{(x-x)^2 + (y-{}^-3)^2}$$
$$(x-0)^2 + (y-3)^2 = (x-x)^2 + (y+3)^2$$

$$x^2 + y^2 - 6y + 9 = 0^2 + y^2 + 6y + 9$$
$$x^2 = 6y + 6y$$
$$x^2 = 12y$$

OR   Focus$(0, p) = (0, 3)$,
$$\therefore \quad p = 3$$
From $x^2 = 4py$
$$x^2 = 4 * 3 * y$$
$$x^2 = 12y$$

(iii) Find the vertex, focus and diretrix of the equation of a parabola $y^2 = -12x$.

Solution

Since the equation is in standard form of $y^2 = {}^-4px$, the vertex is at origin $(0, 0)$

when you compare the two equation
$$^-4p = {}^-12$$
$$\therefore p = 3$$
The focus $(^-p, 0) = (^-3, 0)$
The directrix $(x = p)$ is
$$x = 3$$

(iv) Find the vertex, focus and diretrix of the equation of a parabola $x^2 = {}^-12y$.

Solution

Since the equation is in standard form of $x^2 = {}^-4py$, the vertex is at origin $(0, 0)$

When you compare the two equations
$$^-4p = {}^-12$$
$$\therefore \quad p = 3$$
The focus $(0, {}^-p) = (0, {}^-3)$
The directrix $(y = p)$ is
$$y = 3$$

(v) Find the equation of a parabola opening right with vertex (3, 2) and the focus (6, 2).

Solution

Let $P(x, y)$ be any point on a parabola

the vertex is (3, 2) and the focus F is (6, 2),

Compare with vertex (h, k) open right and F $(h+p, k)$

$$h = 3, k = 2,$$
$$h+p = 6$$
$$3 + p = 6$$
$$p = 3$$

From $(y - k)^2 = 4p(x - h)$

$$(y - 2)^2 = 4*3(x - 3)$$
$$(y - 2)^2 = 12(x - 3)$$

OR Since the vertex is (3, 2) and the focus F is (6, 2),

Compare with vertex (h, k) open right and F $(h+p, k)$

$$h = 3, k = 2,$$
$$h+p = 6$$
$$3 + p = 6$$
$$p = 3$$

directrix $(x = h - p)$ is

$$x = 3 - 3$$
$$x = 0$$

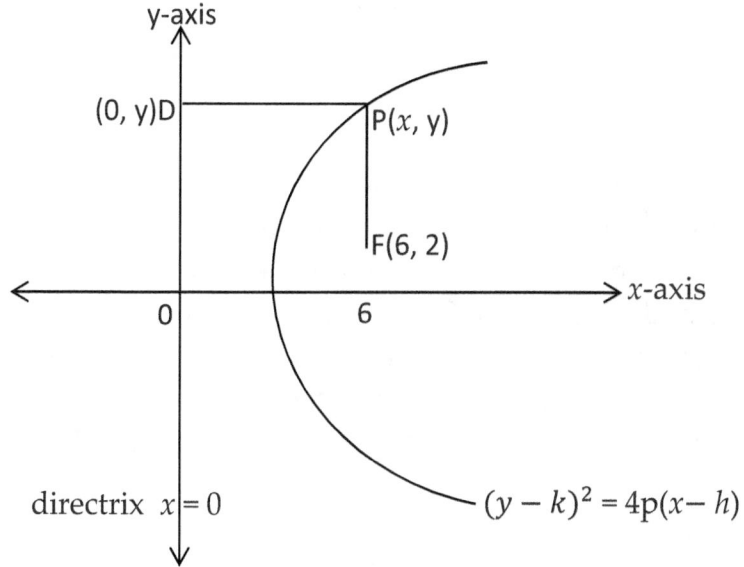

$$\text{From} \quad PF = PD$$
$$\sqrt{(x-6)^2 + (y-2)^2} = \sqrt{(x-0)^2 + (y-y)^2}$$
$$(x-6)^2 + (y-2)^2 = (x-0)^2 + (y-y)^2$$
$$x^2 - 12x + 36 + (y-2)^2 = x^2 + 0$$
$$(y-2)^2 = 12x - 36$$
$$(y-2)^2 = 12(x-3)$$

(vi) Find the vertex, focus and diretrix of the equation of a parabola $(x - 2)^2 = 12(y-3)$

Solution

$$\text{Compare } (x - 2)^2 = 12(y-3) \text{ with}$$
$$(x-h)^2 = 4p(y-k)$$
$$h = 2, \quad k = 3,$$
$$4p = 12$$
$$p = 3$$
$$\text{vertex } (h, k) = (2, 3)$$
$$\text{focus } (h, k + p) = (2, 3 + 3)$$
$$= (2, 6)$$

directrix $(y = k - p)$ is
$$y = 3 - 3$$
$$y = 0$$

Ellipse is a set of all points such that the sum of the distances from each two given fixed point (foci) is constant. F_1 *and* F_2 are focus, and P is any point on the ellipse

Focal radii $PF_1 + PF_2 = $ constant

<u>Center</u> is a midpoint of the line between foci F_1 and F_2

<u>Focal length</u> is a distance from center to focus F_1 *or* F_2

<u>Major axis</u> is a line through center and foci F_1 and F_2

<u>Minor axis</u> is a line through center and perpendicular to F_1 *and* F_2

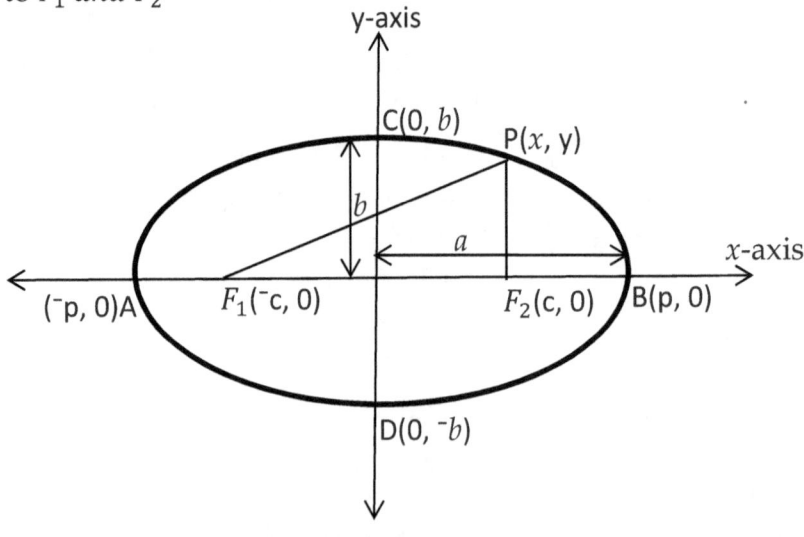

Center $(0, 0)$
$$PF_1 + PF_2 = 2a$$
vertex $(a, 0)$ and $(^-a, 0)$
also co-vertex $(0, b)$ and $(0, ^-b)$

Any given ellipse with center at origin $(0, 0)$ and foci at $(c, 0)$ and $(^-c, 0)$ has an equation in standard form
$$\frac{x^2}{a^2} + \frac{y^2}{b^2} = 1$$

where $a > b$

and $c^2 = a^2 - b^2$

Also any given ellipse with center at origin (0, 0) and foci at (0, c) and (0, ‾c) has an equation in standard form

$$\frac{x^2}{b^2} + \frac{y^2}{a^2} = 1$$

where $a > b$

and $c^2 = a^2 - b^2$

Examples 15.3

C(i) Sketch and find the center and the ends of the axes of

ellipse $\dfrac{x^2}{16} + \dfrac{y^2}{4} = 1$.

Solution

Find intercepts of ellipse,

when $x = 0$

$$\frac{x^2}{16} + \frac{y^2}{4} = 1$$

$$\frac{0^2}{16} + \frac{y^2}{4} = 1$$

$$\frac{y^2}{4} = 1$$

$$y^2 = 4$$

$$\sqrt{y^2} = \sqrt{4}$$

$$y = \pm 2$$

y-intercepts (0, 2) and (0, ‾2)

when $y = 0$

$$\frac{x^2}{16} + \frac{y^2}{4} = 1$$

$$\frac{x^2}{16} + \frac{0^2}{4} = 1$$

$$\frac{x^2}{16} = 1$$
$$x^2 = 16$$
$$\sqrt{x^2} = \sqrt{16}$$
$$x = \pm 4$$

x-intercepts $(4, 0)$ and $(^-4, 0)$

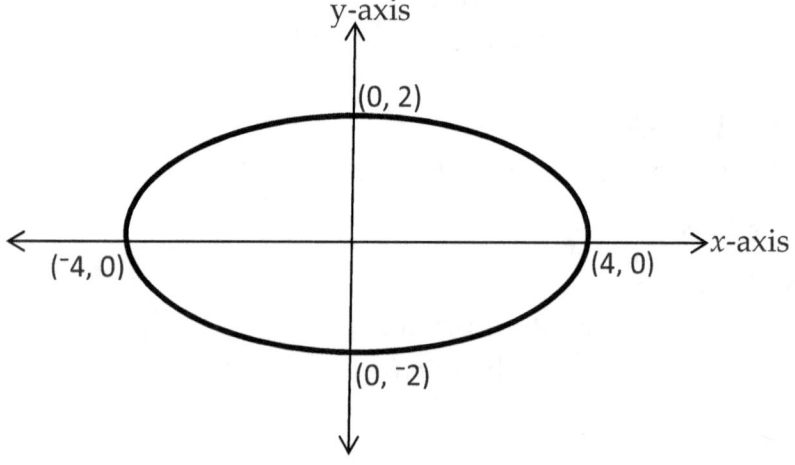

Center at origin $(0, 0)$

(ii) Find the equation of an ellipse, its center and the vertex given the foci $(4, 0)$ and $(^-4, 0)$, the sum of focal radii is *10*

Solution

$$PF_1 + PF_2 = 10$$
$$\sqrt{(x-4)^2 + (y-0)^2} + \sqrt{(x+4)^2 + (y-0)^2} = 10$$
$$\sqrt{x^2 - 8x + 16 + y^2} + \sqrt{x^2 + 8x + 16 + y^2} = 10$$
$$\sqrt{x^2 - 8x + 16 + y^2} = 10 - \sqrt{x^2 + 8x + 16 + y^2}$$

Square both sides

$$x^2 - 8x + 16 + y^2 = 100 - 20\sqrt{x^2 + 8x + 16 + y^2} + x^2 + 8x + 16 + y^2$$

$$^-8x = 100 - 20\sqrt{x^2 + 8x + 16 + y^2} + 8x$$
$$20\sqrt{x^2 + 8x + 16 + y^2} = 8x + 8x + 100$$
$$20\sqrt{x^2 + 8x + 16 + y^2} = 16x + 100$$

Square both sides

$$400(x^2 + 8x + 16 + y^2) = 256x^2 + 3200x + 10000$$

$$400x^2 + 3200x + 6400 + 400y^2 = 256x^2 + 3200x + 10000$$

$$400x^2 - 256x^2 + 400y^2 = 10000 - 6400$$

$$144x^2 + 400y^2 = 3600$$

$$\frac{144x^2}{3600} + \frac{400y^2}{3600} = \frac{3600}{3600}$$

$$\frac{x^2}{25} + \frac{y^2}{9} = 1$$

$$c = 4$$

$$2a = 10$$

$$a = 5$$

Center $(0, 0)$

Vertex $(a, 0)$ and $(^-a, 0) = (5, 0)$ and $(^-5, 0)$

OR

$$c = 4$$

$$2a = 10$$

$$a = 5$$

$$c^2 = a^2 - b^2$$

$$4^2 = 5^2 - b^2$$

$$16 = 25 - b^2$$

$$b^2 = 25 - 16$$

$$b^2 = 9$$

$$b = 3$$

From $\dfrac{x^2}{a^2} + \dfrac{y^2}{b^2} = 1$

$$\frac{x^2}{5^2} + \frac{y^2}{3^2} = 1$$

$$\frac{x^2}{25} + \frac{y^2}{9} = 1$$

Center $(0, 0)$

Vertex $(a, 0)$ and $(^-a, 0) = (5, 0)$ and $(^-5, 0)$

An ellipse with center at *(h, k)* has equation in standard form.

$$\frac{(x-h)^2}{q} + \frac{(y-k)^2}{r} = 1$$

where q and r are different positive numbers

Examples 15.3

c(iii) A graph of $4x^2 + 3y^2 + 24x - 12y + 36 = 0$ is an ellipse with center ($^-$3, 2) and axes along $x = ^-3$ and $y = 2$. Put it into a standard form.

Solution

$$4x^2 + 3y^2 + 24x - 12y + 36 = 0$$
$$4x^2 + 24x + 3y^2 - 12y + 36 = 0$$
$$4(x^2 + 6x) + 3(y^2 - 4y) + 36 = 0$$
$$4[(x + 3)^2 - 3^2] + 3[(y - 2)^2 - 2^2] + 36 = 0$$
$$4[(x + 3)^2 - 9] + 3[(y - 2)^2 - 4] + 36 = 0$$
$$4(x + 3)^2 - 36 + 3(y - 2)^2 - 12 + 36 = 0$$
$$4(x + 3)^2 + 3(y - 2)^2 - 12 + 36 - 36 = 0$$
$$4(x + 3)^2 + 3(y - 2)^2 - 12 = 0$$
$$4(x + 3)^2 + 3(y - 2)^2 = 12$$
$$\frac{4(x+3)^2}{12} + \frac{3(y-2)^2}{12} = \frac{12}{12}$$

$$\frac{(x+3)^2}{3} + \frac{(y-2)^2}{4} = 1$$

Hyperbola is the set of all points such that the absolute value of the difference of the distance from each of the points to two given fixed points (foci) is a constant.

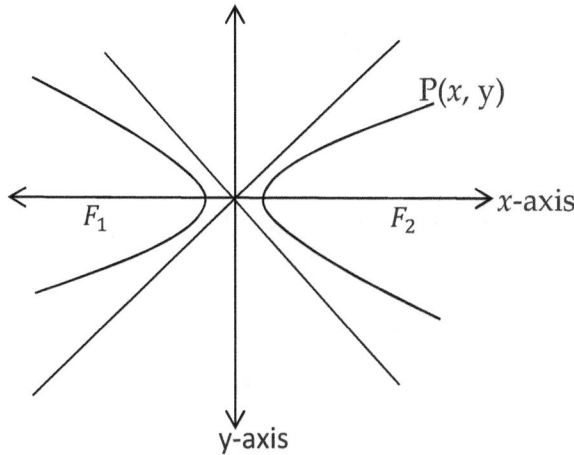

Focus F_1 and F_2

P is any point on hyperbola

$|PF_1 - PF_2| = \text{constant}$

Standard form of hyperbola with center $(0, 0)$, foci $(c, 0)$ and $(^-c, 0)$ is

$$\frac{x^2}{a^2} - \frac{y^2}{b^2} = 1$$

where $c^2 = a^2 + b^2$

$$|PF_1 - PF_2| = 2a$$

<u>Axis</u> is the line through foci F_1 and F_2

<u>Vertices</u> are points where graph crosses axis

<u>Center</u> is a midpoint of line between foci and also midpoint of line between vertices

<u>Asymptotes</u> are $y = \frac{b}{a}x$ and $y = \frac{^-b}{a}x$

Other standard forms of hyperbolas

$$\frac{y^2}{a^2} - \frac{x^2}{b^2} = 1$$

foci at $(0, c)$ and $(0, ^-c)$

Center $(0, 0)$

$$c^2 = a^2 + b^2$$

Asymptotes are $y = \pm \frac{a}{b}x$

$$\frac{(x-h)^2}{a^2} - \frac{(y-k)^2}{b^2} = 1$$

foci at (c, 0) and ($^-$c, 0)

Center (h, k)

$$c^2 = a^2 + b^2$$

Asymptotes are $y-k = \pm \frac{b}{a}(x-h)$

$$\frac{(y-k)^2}{a^2} - \frac{(x-h)^2}{b^2} = 1$$

foci at (0, c) and (0, $^-$c)

Center (h, k)

$$c^2 = a^2 + b^2$$

Asymptotes are $y-k = \pm \frac{a}{b}(x-h)$

Examples 15.3

d(i) Find the equation of hyperbola and its asymptotes with center at *(0, 0)*, foci F_1 *and* F_2 at *(5, 0)* and *($^-$5, 0)* and also with $|PF_1 - PF_2| = 8$

Solution

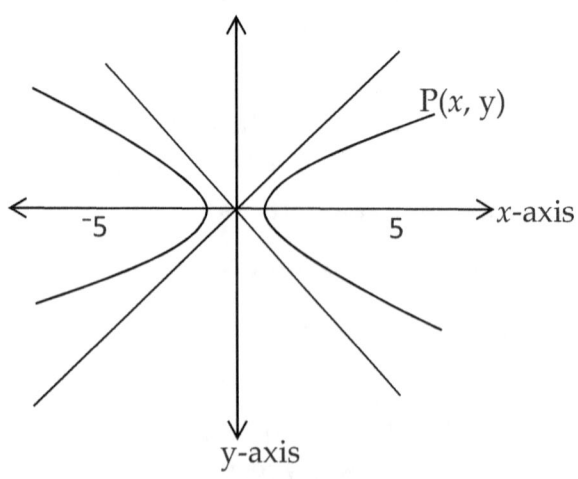

center (0, 0)

F_1 and F_2 at (5, 0) and ($^-$5, 0)

$$c = 5$$

From $|PF_1 - PF_2| = 8$

$$2a = 8$$

$$a = \frac{8}{2}$$

$$a = 4$$

From $c^2 = a^2 + b^2$

$$5^2 = 4^2 + b^2$$

$$b^2 = 5^2 - 4^2$$

$$b^2 = 25 - 16$$

$$b^2 = 9$$

$$b = 3$$

From $\quad \dfrac{x^2}{a^2} - \dfrac{y^2}{b^2} = 1$

$$\frac{x^2}{4^2} - \frac{y^2}{3^2} = 1$$

$$\frac{x^2}{16} - \frac{y^2}{9} = 1$$

Asymptotes are $y = \dfrac{b}{a}x$ and $y = \dfrac{^-b}{a}x$

$$y = \frac{3}{4}x \text{ and } y = \frac{^-3}{4}x$$

Asymptotes are $y = \pm\, \dfrac{3}{4}x$

OR   From $|PF_1 - PF_2| = 8$

$$|\sqrt{(x-5)^2 + y^2} - \sqrt{(x+5)^2 + y^2}| = 8$$

$$\sqrt{x^2 - 10x + 25 + y^2} - \sqrt{x^2 + 10x + 25 + y^2} = \pm 8$$

$$\sqrt{x^2 - 10x + 25 + y^2} = \pm 8 + \sqrt{x^2 + 10x + 25 + y^2}$$

Square both sides

$$x^2 - 10x + 25 + y^2 = 64 \pm 16\sqrt{x^2 + 10x + 25 + y^2} + x^2$$
$$+10x + 25 + y^2$$

$$-10x = 64 \pm 16\sqrt{x^2 + 10x + 25 + y^2} + 10x$$

$$\pm 16\sqrt{x^2 + 10x + 25 + y^2} = 10x + 10x + 64$$
$$\pm 16\sqrt{x^2 + 10x + 25 + y^2} = 20x + 64$$

Square both sides

$$256(x^2 + 10x + 25 + y^2) = 400x^2 + 2560x + 4096$$
$$256x^2 + 2560x + 6400 + 256y^2 = 400x^2 + 2560x + 4096$$
$$256x^2 - 400x^2 + 256y^2 = 4096 - 6400$$
$$^-144x^2 + 256y^2 = {}^-2304$$

Multiply both sides by $^-1$

$$144x^2 - 256y^2 = 2304$$
$$\frac{144x^2}{2304} - \frac{256y^2}{2304} = \frac{2304}{2304}$$

$$\frac{x^2}{16} - \frac{y^2}{9} = 1$$

Asymptotes are $y = \dfrac{b}{a}x$ and $y = \dfrac{^-b}{a}x$

$$y = \frac{3}{4}x \quad \text{and} \quad y = \frac{^-3}{4}x$$

Asymptotes are $y = \pm \dfrac{3}{4}x$

(ii) Find the center, vertices and asymptotes of the
hyperbola $\dfrac{y^2}{25} - \dfrac{x^2}{16} = 1$

Solution

$$\frac{y^2}{25} - \frac{x^2}{16} = 1$$

is in standard form where the center at $(0, 0)$

When $x = 0$

$$\frac{y^2}{25} - \frac{0^2}{16} = 1$$

$$\frac{y^2}{25} = 1$$
$$y^2 = 25$$

$$y = 5 \quad \text{or} \quad y = {}^-5$$

$$y\text{-intercepts (vertices)} = (0, 5) \quad \text{and} \quad (0, {}^-5)$$

$$\text{when } y = 0$$

$$\frac{0^2}{25} - \frac{x^2}{16} = 1$$

$$-\frac{x^2}{16} = 1$$

$$x^2 = {}^-16 \quad \text{No real solution}$$

$$\frac{y^2}{25} - \frac{x^2}{16} = 1$$

$$\frac{y^2}{5^2} - \frac{x^2}{4^2} = 1$$

Compare with $\dfrac{y^2}{a^2} - \dfrac{x^2}{b^2} = 1$

$$a = 5 \quad \text{and} \quad b = 4$$

Asymptotes are $y = \dfrac{a}{b}x$ and $y = \dfrac{{}^-a}{b}x$

$$y = \frac{5}{4}x \quad \text{and} \quad y = \frac{{}^-5}{4}x$$

(iii) Write the equation $y^2 - x^2 - 10y - 4x + 12 = 0$ in standard form of hyperbola and find its equation of asymptotes.

Solution

$$y^2 - x^2 - 10y - 4x + 12 = 0 \quad \text{by completing squares}$$

$$y^2 - 10y - x^2 - 4x + 12 = 0$$

$$y^2 - 10y + \left(\frac{10}{2}\right)^2 - x^2 - 4x + \left(\frac{4}{2}\right)^2 + 12 = 0 + \left(\frac{10}{2}\right)^2 + \left(\frac{4}{2}\right)^2$$

$$y^2 - 10y + 5^2 - x^2 - 4x + 2^2 + 12 = 0 + 5^2 + 2^2$$

$$y^2 - 10y + 25 - x^2 - 4x + 4 + 12 = 0 + 25 + 4$$

$$y^2 - 10y + 25 - x^2 - 4x + 16 = 29$$

$$y^2 - 10y + 25 - x^2 - 4x + 16 - 16 = 29 - 16$$
$$y^2 - 10y + 25 - x^2 - 4x - 4 = 13 - 4$$
$$y^2 - 10y + 25 - (x^2 + 4x + 4) = 9$$
$$(y - 5)^2 - (x + 2)^2 = 9$$
$$\frac{(y-5)^2}{9} - \frac{(x+2)^2}{9} = \frac{9}{9}$$

$$\frac{(y-5)^2}{9} - \frac{(x+2)^2}{9} = 1$$

$$\frac{(y-5)^2}{3^2} - \frac{(x+2)^2}{3^2} = 1 \qquad \text{compare with}$$

$$\frac{(y-k)^2}{a^2} - \frac{(x-h)^2}{b^2} = 1$$

$$k = 5 \qquad h = {}^-2, \qquad a = b = 3$$

when $a = b$, the hyperbola is called <u>equilateral hyperbola.</u>

equations of asymptotes are

$$y - k = \pm \frac{a}{b}(x - h)$$

$$y - 5 = \pm \frac{3}{3}(x - {}^-2)$$

$$y - 5 = \pm(x + 2)$$

Exercise 15A

(1) What is the name of an *8* sided polygon?

(2) What kind of angles are X, Y and Z?

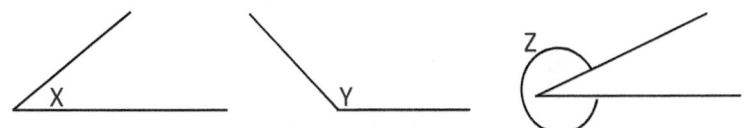

(3) What is the difference between an equilateral polygon and an equiangular polygon?

(4)

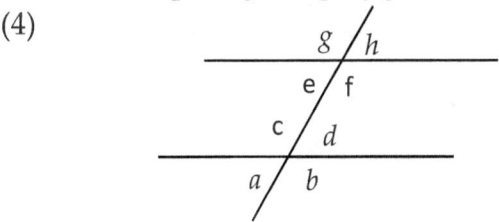

(a) Given $a° = 45°$; find all other angles and their relationship between each other.

(b) Which angles are congruent angles?

5(a) Find an exterior angle of a regular nonagon.

(b) Find an exterior angle of a regular decagon.

(c) Find an exterior angle of a regular hectagon.

(6) The length of a regular heptagon is *5m* and its apothem is *8m*. Find the area of the regular heptagon.

(7) The side of a regular pentagon is *8m* and the radius is *5m*. Find the area of a regular pentagon.

(8) The side of a regular pentagon is *12m* and the radius is *10m*. Find the area of a regular pentagon.

(9) Find the perimeter of an equilateral triangle with side measure *10.7in*.

(10) The vertex angle of an isosceles triangle is *40.2°* Find the base angle.

(11) Find the perimeter and the area of a triangle bellow.

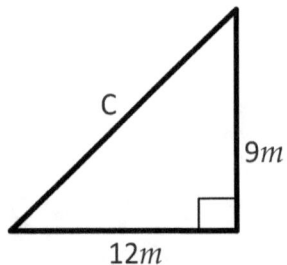

(12) The triangles area similar and the sides are proportion Find b.

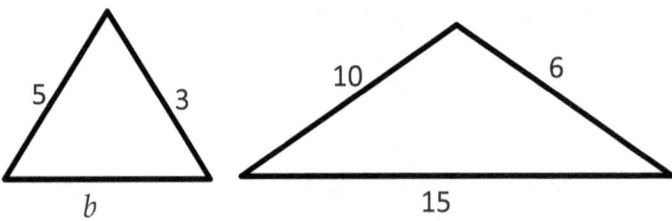

(13) Two triangles said to be Congruent. Find angle b

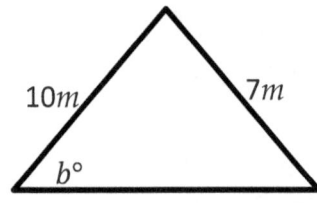 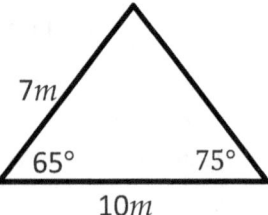

(14) The perimeter of a rectangle is *38m*, its length is *12m* and its width is *3x*. Find x

(15) The length of a rectangle is twice the width. Find the width given the area of a rectangle is *242m²*.

(16) The length of a rectangle is nine more than the width. Find width given the perimeter of the rectangle is *54m*.

(17) Find the area of the following figures:
Given *a = 7m, b = 11m, h = 6m* and diameter *D = 8m*

(a)

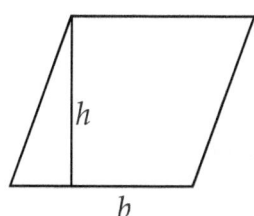

(b)

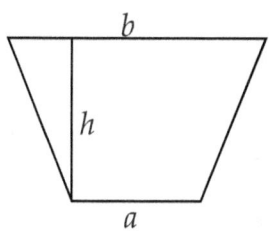

(c)

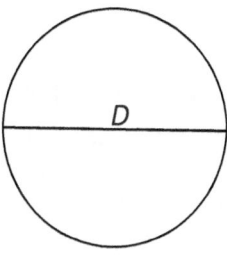

(18) Find the volume of figures given $w = 2.5m$, $L = 12.8m$, $h = 5m$ and for cylinder and cone diameter $D = 10m$.

(a)

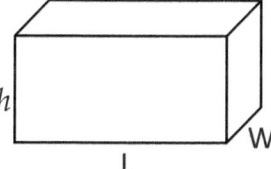

(b)

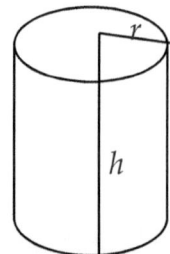

(c)

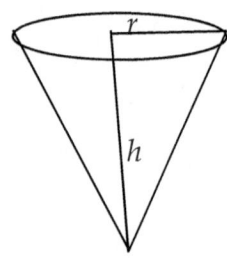

(19) The following figures are filled with water.
Given the diameter of the cylinder and cone $D = 12m$,
$h = 8m$, $h_1 = 4m$, $h_2 = 3m$ and $r_2 = 2m$.   Find the
volume of the empty space that is not filled with water.

(a)

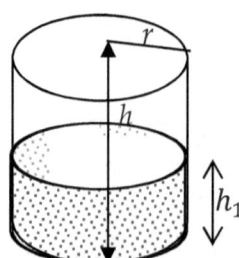

(b)

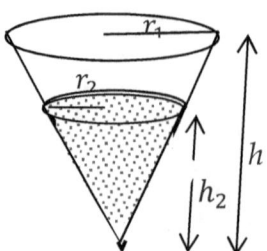

(20) The vertex angle of an isosceles triangle exceeds the measurement of each base angle by *30*°. Find the value of vertex angle of the triangle.

(21) Find a perimeter of a triangle below

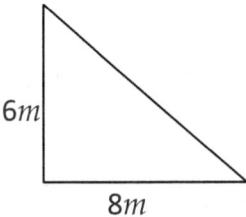

(22) Find the arc length *AB* with radius *BC* = *9m*

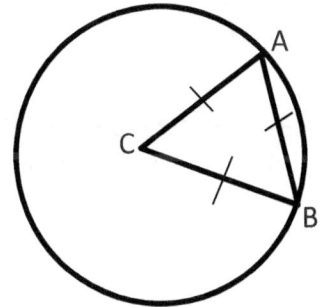

(23) Find the equation of the circle with center at origin and radius *13*.

(24) Find the center and radius of the equation of a circle $x^2 + y^2 = 121$.

(25) Find equation of the circle with center at (2, ⁻5) and with radius *9*.

(26) Find the center and the radius of the circle with equation $3x^2 + 3y^2 + 12x - 18y + 9 = 18$.

(27) Find the equation of a parabola with vertex at origin and focus (5, 0).

(28) Find the equation of a parabola with vertex at origin and focus (0, 7).

(29) Find the vertex, focus and diretrix of the equation of a parabola $y^2 = {}^-52x$.

(30) Find the vertex, focus and diretrix of the equation of a parabola $x^2 = {}^-24y$.

(31) Find the equation of a parabola opening right with vertex $(5, 3)$ and the focus $(9, 3)$.

(32) Find the vertex, focus and diretrix of the equation of a parabola $(x - 7)^2 = 44(y-2)$

(33) Sketch and find the center and the ends of the axes of ellipse $\dfrac{x^2}{36} + \dfrac{y^2}{9} = 1$.

(34) Find the equation of an ellipse, its center and the vertex given the foci $(8, 0)$ and $({}^-8, 0)$, the sum of focal radii is 20

(35) A graph of $3x^2 + 4y^2 + 24x - 24y + 72 = 0$ is an ellipse with center $({}^-4, 3)$ and axes along $x = {}^-4$ and $y = 3$ put it into a standard form.

(36) Find the equation of hyperbola and its asymptotes with center at $(0, 0)$, foci F_1 and F_2 at $(10, 0)$ and $({}^-10, 0)$ and also with $|PF_1 - PF_2| = 16$

(37) Find the center, vertices and asymptotes of the hyperbola $\dfrac{y^2}{144} - \dfrac{x^2}{81} = 1$.

(38) Write the equation $y^2 - x^2 - 4y - 6x - 41 = 0$ in standard form of hyperbola and find its equation of asymptotes.

(39) Sketch the graph of $x^2 = 8(y + 2)$ and find the vertex and the intercept(s).

(40) Find the equation of an ellipse, its center and the vertex given the foci $(0, 4)$ and $(0, {}^-4)$, the sum of focal radii is 10.

(41) Find the equation of hyperbola and its asymptotes with center at $(0, 0)$, foci F_1 and F_2 at $(0, 5)$ and $(0, {}^-5)$ and also with $|PF_1 - PF_2| = 8$

(42) Write the equation $4x^2 - 9y^2 - 24x - 36y - 36 = 0$ in standard form of hyperbola and find its equation of asymptotes.

SOLUTIONS FOR EXERCISE 15A

(1) Octagon

(2) X is acute angle

Y is obtuse angle

Z is reflex angle

(3) Equilateral polygon is a polygon with all sides equal while an equiangular polygon is a polygon with all angles equal.

(4)

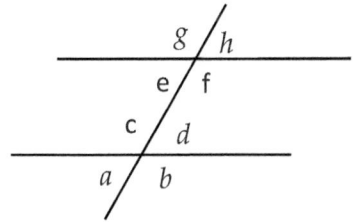

(a) $a° = d° = 45°$ Vertical angles

$b° = 180° - a°$ Straight angles(angles on line=180°)

$b° = 180° - 45°$

$b° = 135°$

$b° = c° = 135°$ Vertical angles

$d° = e° = 45°$ Alternate interior angles

$b° = g° = 135°$ Alternate exterior angles

$g° = f° = 135°$ Vertical angles

$h° = 180° - f°$ Straight angles(angles on line=180°)

$h° = 180° - 135°$

$h° = 45°$

(b) $a°$, $d°$, $e°$ and $h°$ are congruent angles also
        $b°$, $c°$, $f°$ and $g°$ are congruent angles

5(a) Exterior angle of a regular nonagon $= \dfrac{360°}{9} = 40°$

(b) Exterior angle of a regular decagon $= \dfrac{360°}{10} = 36°$

(c) Exterior angle of a regular hectagon $= \dfrac{360°}{100} = 3.6°$

(6)
$$\text{Area} = \frac{1}{2}ap$$
But perimeter $P = 5m \times 7$ sides
$$= 35m$$
$$\text{Area} = \frac{1}{2} \times 8m \times 35m$$
$$= 4m \times 35m$$
$$= 140m^2$$

(7)

From Pythagorean theory
$$r^2 = a^2 + 4^2$$
$$5^2 = a^2 + 4^2$$
$$25 = a^2 + 16$$
$$25 - 16 = a^2$$
$$9 = a^2$$
$$3 = a$$
∴ apothem $(a) = 3m$

Perimeter $P = 8m + 8m + 8m + 8m + 8m$

$$= 40m$$

From Area $= \frac{1}{2}ap$

$$= \frac{1}{2} \times 3m \times 40m$$

Area $= 60m^2$

(8)

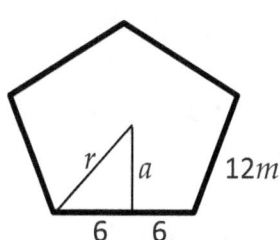

From Pythagorean theory

$$r^2 = a^2 + 6^2$$

$$10^2 = a^2 + 6^2$$

$$100 = a^2 + 36$$

$$100 - 36 = a^2$$

$$64 = a^2$$

$$8 = a$$

$\therefore$ apothem $(a) = 8m$

Perimeter $P = 12m + 12m + 12m + 12m + 12m$

$$= 60m$$

From Area $= \frac{1}{2}ap$

$$= \frac{1}{2} \times 8m \times 60m$$

Area $= 240m^2$

(9) For equilateral triangle all sides are equal

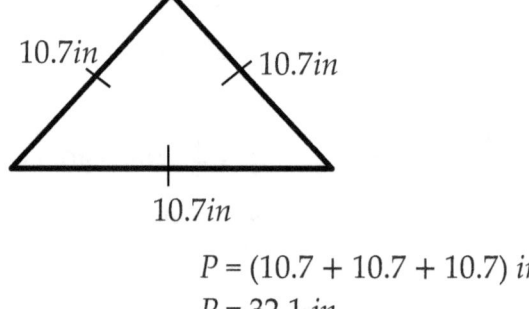

$$P = (10.7 + 10.7 + 10.7)\ in$$
$$P = 32.1\ in$$

(10)   Let the base angle of an isosceles triangle be x

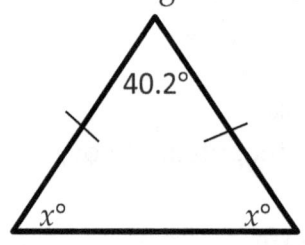

$$x + x + 40.2° = 180°$$
$$2x + 40.2° = 180°$$
$$2x + 40.2° - 40.2° = 180° - 40.2°$$
$$2x = 139.8°$$
$$\frac{2x}{2} = \frac{139.8°}{2}$$
$$x = 69.9°$$

∴ the base angle is 69.9°

(11)   Let the hypotenuse of a triangle be c

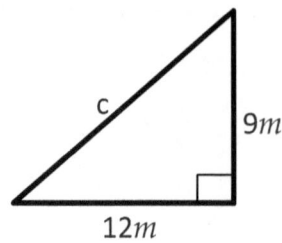

From Pythagorean theory

$$c^2 = 12^2 + 9^2$$
$$c^2 = 144 + 81$$
$$c^2 = 225$$
$$c = 15m$$

The perimeter $= 15m + 12m + 9m$
$$= 36m$$

$$\text{Area} = \frac{1}{2}bh$$

$$= \frac{1}{2} \times 12m \times 9m$$
$$= 54m^2$$

(12)

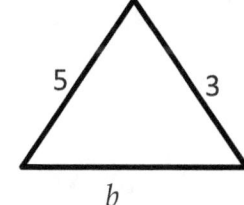 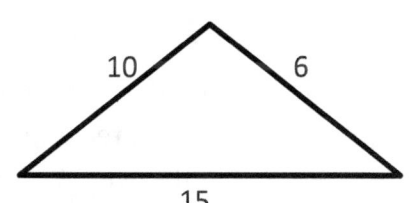

$$3 \times 2 = 6$$
$$5 \times 2 = 10$$
$$b \times 2 = 15$$
$$b = \frac{15}{2}$$
$$\therefore \quad b = 7.5$$

(13)

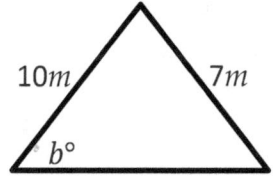 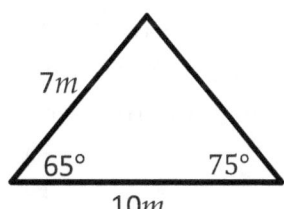

$$b = 75°$$

(14)

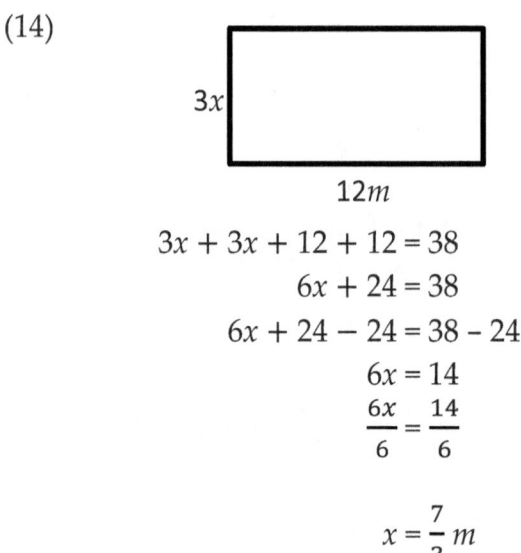

$$3x + 3x + 12 + 12 = 38$$
$$6x + 24 = 38$$
$$6x + 24 - 24 = 38 - 24$$
$$6x = 14$$
$$\frac{6x}{6} = \frac{14}{6}$$

$$x = \frac{7}{3}m$$
$$= 2.333\ m$$

(15) Let the width of a rectangle be w
The length of a rectangle is $2w$
From the area $A = LW$
$$242 = 2w \times w$$
$$242 = 2w^2$$
$$\frac{242}{2} = \frac{2w^2}{2}$$
$$121 = w^2$$
$$w = 11$$
∴ the width of a rectangle is $11m$

(16) Let the width of a rectangle be x
The length of the rectangle is $x + 9$
From Perimeter $P = 2L + 2W$
$$54 = 2(x + 9) + 2x$$
$$54 = 2x + 18 + 2x$$
$$54 = 4x + 18$$
$$54 - 18 = 4x + 18 - 18$$

$$36 = 4x$$
$$x = \frac{36}{4}$$
$$x = 9$$

∴ the width of a rectangle is $9m$

17a,

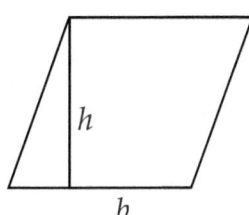

From area $A = bh$
$$A = 11m \times 6m$$
$$A = 66m^2$$

(b)

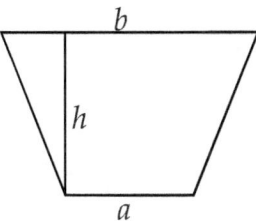

From area $A = \frac{1}{2}(a + b)h$

$$A = \frac{1}{2}(7 + 11)6$$

$$A = \frac{1}{2}(18)6$$

$$= \frac{108}{2}$$
$$A = 54m^2$$

(c)

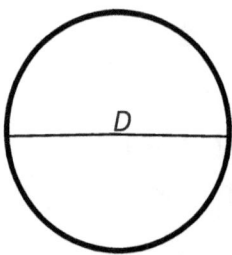

r is the radius

$$r = \frac{D}{2}$$

$$= \frac{8}{2}$$

$$r = 4m$$

From area $A = \pi r^2$

$$A = \pi 4^2$$

$$A = 16\pi m^2$$

$$A = 50.24 m^2$$

18(a)

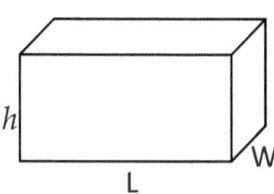

Volume $V = Lwh$

$$= 12.8m \times 2.5m \times 5m$$

$$V = 160m^3$$

(b)

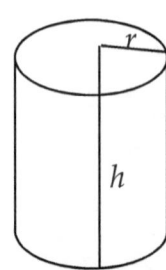

Volume $V = \pi r^2 h$

$$\text{But } r = \frac{D}{2}$$

$$= \frac{10}{2}$$

$$r = 5m$$

$$V = \pi 5^2 * 5$$

$$V = 125\pi m^3$$

$$V = 392.5 m^3$$

(c)

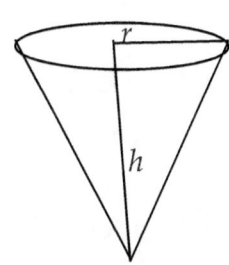

Volume $V = \frac{1}{3} \pi r^2 h$

$$\text{But } r = \frac{D}{2}$$

$$= \frac{10}{2}$$

$$r = 5m$$

$$V = \frac{1}{3} \pi 5^2 * 5$$

$$V = \frac{125\pi}{3} m^3$$

$$V = 130.8 m^3$$

(19) (a)

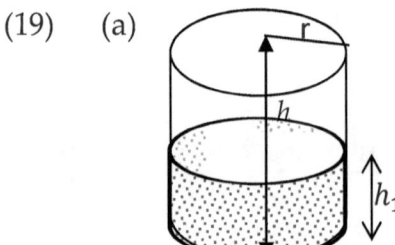

Volume of cylinder $V = \pi r^2 h$
$$\text{But } r = \frac{D}{2}$$

$$= \frac{12}{2}$$
$$r = 6m$$
$$V = \pi 6^2 * 8$$
$$V = 904.32 m^3$$

Volume of water in the cylinder
$$V = \pi r^2 h_1$$

$$\text{But } r = \frac{D}{2}$$
$$= \frac{12}{2}$$
$$r = 6m$$
$$V = \pi 6^2 * 4$$
$$V = 452.16 m^3$$

Volume of empty space in the cylinder
$$904.32 m^3 - 452.16 m^3 = 452.16 m^3$$

Or height of empty space $= h - h_1$
$$8m - 4m = 4m$$
$$V = \pi 6^2 * 4$$
$$V = 452.16 m^3$$

(b)

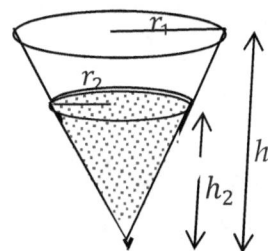

Volume of the cone $V = \dfrac{1}{3}\pi r_1{}^2 h$

But $r_1 = \dfrac{D}{2}$

$$= \dfrac{12}{2}$$
$$r_1 = 6m$$
$$V = \dfrac{1}{3}\pi 6^2 * 8$$

$$V = 301.44m^3$$

Volume of water in the cone

$$V = \dfrac{1}{3}\pi r_2{}^2 h_2$$

$$V = \dfrac{1}{3}\pi 2^2 * 3$$
$$V = 12.56m^3$$

Volume of empty space in the cone

$$301.44m^3 - 12.56m^3 = 288.88m^3$$

(20)

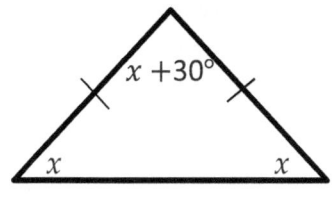

Let the base angle be x
the vertex angle is $(x + 30°)$
$x + x + x + 30° = 180°$
$3x + 30° = 180°$
$3x + 30° - 30° = 180° - 30°$
$3x = 150°$
$x = 50°$
The vertex angle $= 50° + 30°$
$= 80°$

(21)

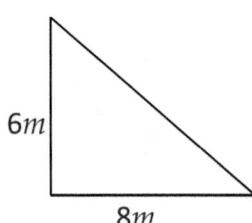

6m

8m

Using a Pythagorean formula
From $c^2 = a^2 + b^2$
$c^2 = 8^2 + 6^2$
$c^2 = 64 + 36$
$c^2 = 100$
$c^2 = 10^2$
$c = 10$
Perimeter $= 10m + 8m + 6m$
$= 24m$

(22) From arc length formula $= \dfrac{n}{360°} 2\pi r$

$= \dfrac{60°}{360°} 2\pi * 9$
$= 3\pi m$
$= 9.42m$

(23)   From $x^2 + y^2 = r^2$
$$x^2 + y^2 = 13^2$$
$$x^2 + y^2 = 169$$

(24)   From $x^2 + y^2 = r^2$
The center is at origin (0, 0).
$$x^2 + y^2 = 121$$
$$x^2 + y^2 = 11^2$$
$$r^2 = 11^2$$
∴ Radius $r = 11$

(25) From $(x - h)^2 + (y - k)^2 = r^2$
$$(x - 2)^2 + (y - {}^-5)^2 = 9^2$$
$$(x - 2)^2 + (y + 5)^2 = 81$$
$$x^2 - 4x + 4 + y^2 + 10y + 25 = 81$$
$$x^2 + y^2 - 4x + 10y + 4 + 25 - 81 = 0$$
$$x^2 + y^2 - 4x + 10y - 52 = 0$$

(26) $3x^2 + 3y^2 + 12x - 18y + 9 = 18.$
$$3x^2 + 3y^2 + 12x - 18y - 9 = 0$$
Divide each term by 3
$$x^2 + y^2 + 4x - 6y - 3 = 0$$
$$x^2 + 4x + y^2 - 6y - 3 = 0$$
Square half of the coefficient of x and y and add it on both sides.
$$x^2 + 4x + \left(\frac{4}{2}\right)^2 + y^2 - 6y + \left(\frac{-6}{2}\right)^2 - 3 = \left(\frac{4}{2}\right)^2 + \left(\frac{-6}{2}\right)^2$$
$$x^2 + 4x + (2)^2 + y^2 - 6y + ({}^-3)^2 - 3 = (2)^2 + ({}^-3)^2$$
$$x^2 + 4x + 4 + y^2 - 6y + 9 - 3 = 4 + 9$$
$$x^2 + 4x + 4 + y^2 - 6y + 9 = 4 + 9 + 3$$
$$(x + 2)^2 + (y - 3)^2 = 16$$
compare with
$$(x - h)^2 + (y - k)^2 = r^2$$
The center $(h, k) = ({}^-2, 3)$
The radius $r^2 = 16$

375

$$r^2 = 4^2$$
$$r = 4$$

(27) Let $P(x, y)$ be any point on a parabola
Since the focus F is $(5, 0)$ and vertex $(0, 0)$,
The directrix D is $x = {}^-5$

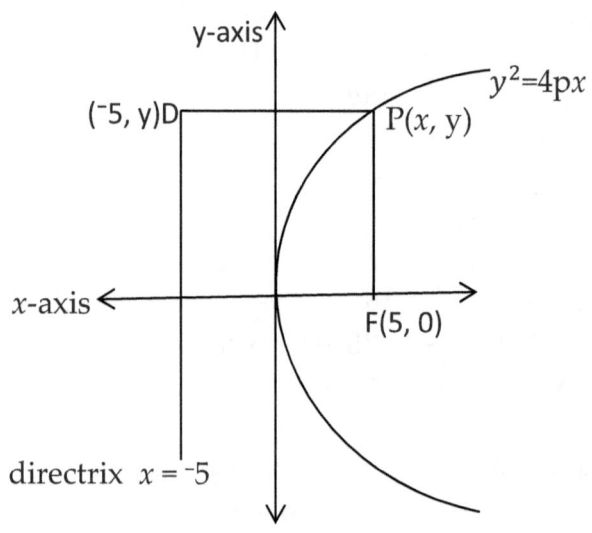

From $PF = PD$

$$\sqrt{(x-5)^2 + (y-0)^2} = \sqrt{(x-{}^-5)^2 + (y-y)^2}$$
$$(x-5)^2 + (y-0)^2 = (x+5)^2 + (y-y)^2$$
$$x^2 - 10x + 25 + y^2 = x^2 + 10x + 25 + 0^2$$
$$y^2 = 10x + 10x$$
$$y^2 = 20x$$

OR   Focus$(p, 0) = (5, 0)$,
$$\therefore p = 5$$
From $y^2 = 4px$
$$y^2 = 4 * 5 * x$$
$$y^2 = 20x$$

(28) Let $P(x, y)$ be any point on a parabola
Since the focus F is $(0, 7)$ and vertex $(0, 0)$,
The directrix D is $y = {}^-7$

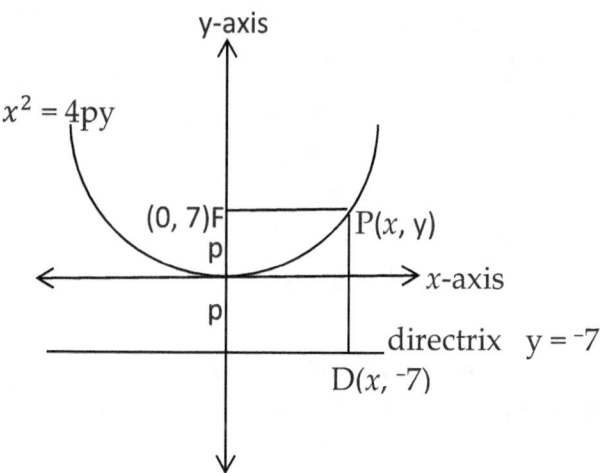

From $PF = PD$

$$\sqrt{(x - 0)^2 + (y - 7)^2} = \sqrt{(x - x)^2 + (y - {}^-7)^2}$$
$$(x - 0)^2 + (y - 7)^2 = (x - x)^2 + (y + 7)^2$$
$$x^2 + y^2 - 14y + 49 = 0^2 + y^2 + 14y + 49$$
$$x^2 = 14y + 14y$$
$$x^2 = 28y$$

OR   Focus$(0, p) = (0, 7)$,
$$\therefore p = 7$$
From $x^2 = 4py$
$$x^2 = 4 * 7 * y$$
$$x^2 = 28y$$

(29) Since the equation is in standard form of $y^2 = {}^-4px$,
$y^2 = {}^-52x$ has its vertex at origin $(0, 0)$.
When you compare the two equations

$$-4p = -52$$
$$\therefore\ p = 13$$
The focus $(-p, 0) = (-13, 0)$
The directrix $(x = p)$ is $x = 13$

(30) Since the equation is in standard form of $x^2 = -4py$,
$x^2 = -4py$ has its vertex at origin $(0, 0)$.
When you compare the two equations
$$-4p = -24$$
$$\therefore p = 6$$
The focus $(0, -p) = (0, -6)$
The directrix $(y = p)$ is $y = 6$

(31) Let $P(x, y)$ be any point on a parabola
the vertex is $(5, 3)$ and the focus F is $(9, 3)$,
Compare with vertex (h, k) open right and F $(h+p, k)$
$$h = 5, k = 3,$$
$$h + p = 9$$
$$5 + p = 9$$
$$p = 4$$
From $(y - k)^2 = 4p(x - h)$
$$(y - 3)^2 = 4* 4(x - 5)$$
$$(y - 3)^2 = 16(x - 5)$$

OR Since the vertex is $(5, 3)$ and the focus F is $(9, 3)$,
Compare with vertex (h, k) open right and F $(h+p, k)$
$$h = 5,\ \ k = 3,$$
$$h + p = 9$$
$$5 + p = 9$$
$$p = 4$$
directrix D $(x = h - p)$
$$x = 5 - 4$$
$$x = 1$$

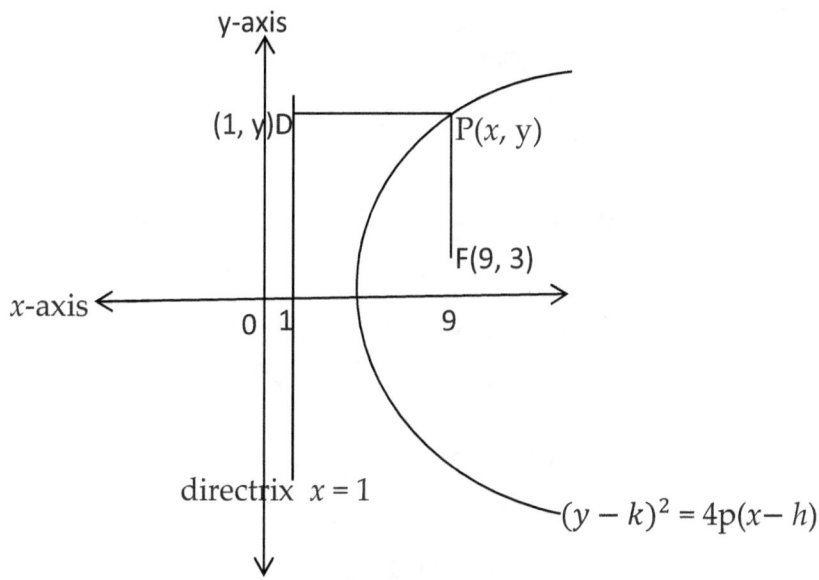

$$\text{From} \quad PF = PD$$
$$\sqrt{(x-9)^2 + (y-3)^2} = \sqrt{(x-1)^2 + (y-y)^2}$$
$$(x-9)^2 + (y-3)^2 = (x-1)^2 + (y-y)^2$$
$$x^2 - 18x + 81 + (y-3)^2 = x^2 - 2x + 1 + 0$$
$$(y-3)^2 = 18x - 2x - 81 + 1$$
$$(y-3)^2 = 16x - 80$$
$$(y-3)^2 = 16(x-5)$$

(32)   Compare $(x-7)^2 = 44(y-2)$ with
$$(x-h)^2 = 4p(y-k)$$
$$h = 7, \quad k = 2,$$
$$4p = 44$$
$$p = 11$$
$$\text{vertex}(h, k) = (7, 2)$$
$$\text{focus}(h, k+p) = (7, 2+11)$$
$$= (7, 13)$$
directrix$(y = k-p)$ is $y = 2-11$
$$y = {}^{-}9$$

(33)

$$\text{when } x = 0$$

$$\frac{x^2}{36} + \frac{y^2}{9} = 1$$

$$\frac{0^2}{36} + \frac{y^2}{9} = 1$$

$$\frac{y^2}{9} = 1$$
$$y^2 = 9$$
$$\sqrt{y^2} = \sqrt{9}$$
$$y = \pm 3$$

y-intercepts $(0, 3)$ and $(0, {}^-3)$

$$\text{when } y = 0$$

$$\frac{x^2}{36} + \frac{y^2}{9} = 1$$

$$\frac{x^2}{36} + \frac{0^2}{9} = 1$$

$$\frac{x^2}{36} = 1$$
$$x^2 = 36$$
$$\sqrt{x^2} = \sqrt{36}$$
$$x = \pm 6$$

x-intercepts $(6, 0)$ and $({}^-6, 0)$

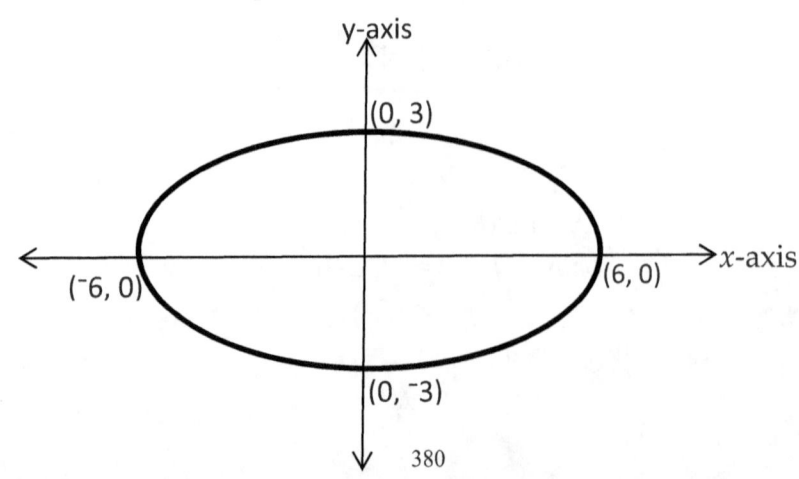

Center at origin (0, 0)

(34)
$$PF_1 + PF_2 = 20$$
$$\sqrt{(x-8)^2 + (y-0)^2} + \sqrt{(x+8)^2 + (y-0)^2} = 20$$
$$\sqrt{x^2 - 16x + 64 + y^2} + \sqrt{x^2 + 16x + 64 + y^2} = 20$$
$$\sqrt{x^2 - 16x + 64 + y^2} = 20 - \sqrt{x^2 + 16x + 64 + y^2}$$

Square both sides

$$x^2 - 16x + 64 + y^2 = 400 - 40\sqrt{x^2 + 16x + 64 + y^2} + x^2 + 16x + 64 + y^2$$

$$^-16x = 400 - 40\sqrt{x^2 + 16x + 64 + y^2} + 16x$$
$$40\sqrt{x^2 + 16x + 64 + y^2} = 16x + 16x + 400$$
$$40\sqrt{x^2 + 16x + 64 + y^2} = 32x + 400$$

Square both sides

$$1600(x^2 + 16x + 64 + y^2) = 1024x^2 + 25600x + 160000$$
$$1600x^2 + 25600x + 102400 + 1600y^2 = 1024x^2 + 25600x + 160000$$

$$1600x^2 - 1024x^2 + 1600y^2 = 160000 - 102400$$
$$576x^2 + 1600y^2 = 57600$$
$$\frac{576x^2}{57600} + \frac{1600y^2}{57600} = \frac{57600}{57600}$$
$$\frac{x^2}{100} + \frac{y^2}{36} = 1$$
$$c = 8$$
$$2a = 20,$$
$$a = 10$$

Center (0, 0)

Vertex $(a, 0)$ and $(^-a, 0) = (10, 0)$ and $(^-10, 0)$

OR
$$c = 8$$
$$2a = 20,$$
$$a = 10$$
$$c^2 = a^2 - b^2$$
$$8^2 = 10^2 - b^2$$

$$64 = 100 - b^2$$
$$b^2 = 100 - 64$$
$$b^2 = 36$$
$$b^2 = 6^2$$
$$b = 6$$

From $\dfrac{x^2}{a^2} + \dfrac{y^2}{b^2} = 1$

$$\dfrac{x^2}{10^2} + \dfrac{y^2}{6^2} = 1$$

$$\dfrac{x^2}{100} + \dfrac{y^2}{36} = 1$$

Center $(0, 0)$

Vertex $(a, 0)$ and $(^-a, 0) = (10, 0)$ and $(^-10, 0)$

(35) $3x^2 + 4y^2 + 24x - 24y + 72 = 0$
$3x^2 + 24x + 4y^2 - 24y + 72 = 0$
$3(x^2 + 8x) + 4(y^2 - 6y) + 72 = 0$
$3[(x + 4)^2 - 4^2] + 4[(y - 3)^2 - 3^2] + 72 = 0$
$3[(x + 4)^2 - 16] + 4[(y - 3)^2 - 9] + 72 = 0$
$3(x + 4)^2 - 48 + 4(y - 3)^2 - 36 + 72 = 0$
$3(x + 4)^2 + 4(y - 3)^2 - 48 + 72 - 36 = 0$
$3(x + 4)^2 + 4(y - 3)^2 - 12 = 0$
$3(x + 4)^2 + 4(y - 3)^2 = 12$

$$\dfrac{3(x+4)^2}{12} + \dfrac{4(y-3)^2}{12} = \dfrac{12}{12}$$

$$\dfrac{(x+4)^2}{4} + \dfrac{(y-3)^2}{3} = 1$$

(36)

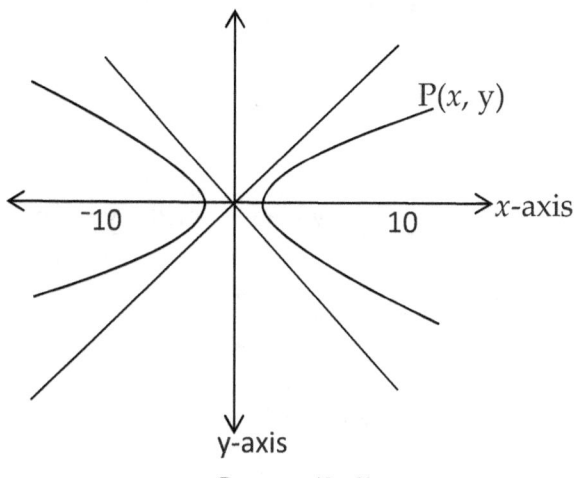

Center $(0, 0)$

$P(x, y)$ F_1 and F_2 at $(10, 0)$ and $(^-10, 0)$

$$c = 10$$

From $|PF_1 - PF_2| = 16$

$$2a = 16$$

$$a = \frac{16}{2}$$

$$a = 8$$

From $c^2 = a^2 + b^2$

$$10^2 = 8^2 + b^2$$

$$b^2 = 10^2 - 8^2$$

$$b^2 = 100 - 64$$

$$b^2 = 36$$

$$b^2 = 6^2$$

$$b = 6$$

From $\dfrac{x^2}{a^2} - \dfrac{y^2}{b^2} = 1$

$$\frac{x^2}{8^2} - \frac{y^2}{6^2} = 1$$

$$\frac{x^2}{64} - \frac{y^2}{36} = 1$$

Asymptotes are $\quad y = \dfrac{b}{a}x \quad$ and $\quad y = \dfrac{^-b}{a}x$

$$y = \dfrac{6}{8}x \quad \text{and} \quad y = \dfrac{^-6}{8}x$$

$$y = \dfrac{3}{4}x \quad \text{and} \quad y = \dfrac{^-3}{4}x$$

Asymptotes are $\quad y = \pm\dfrac{3}{4}x$

OR $\qquad$ From $|PF_1 - PF_2| = 16$

$|\sqrt{(x-10)^2 + (y-0)^2} - \sqrt{(x - {}^-10)^2 + (y-0)^2}| = 16$

$|\sqrt{(x-10)^2 + y^2} - \sqrt{(x+10)^2 + y^2}| = 16$

$\sqrt{x^2 - 20x + 100 + y^2} - \sqrt{x^2 + 20x + 100 + y^2} = \pm 16$

$\sqrt{x^2 - 20x + 100 + y^2} = \pm 16 + \sqrt{x^2 + 20x + 100 + y^2}$

Square both sides

$x^2 - 20x + 100 + y^2 = 256 \pm 32\sqrt{x^2 + 20x + 100 + y^2} + x^2 + 20x + 100 + y^2$

$-20x = 256 \pm 32\sqrt{x^2 + 20x + 100 + y^2} + 20x$

$\pm 32\sqrt{x^2 + 20x + 100 + y^2} = 20x + 20x + 256$

$\pm 32\sqrt{x^2 + 20x + 100 + y^2} = 40x + 256$

Square both sides

$1024(x^2 + 20x + 100 + y^2) = 1600x^2 + 20480x + 65536$

$1024x^2 + 20480x + 102400 + 1024y^2 = 1600x^2 + 20480x + 65536$

$1024x^2 - 1600x^2 + 1024y^2 = 65536 - 102400$

$^-576x^2 + 1024y^2 = {}^-36864$

Multiply by $^-1$ on both sides

$576x^2 - 1024y^2 = 36864$

$\dfrac{576x^2}{36864} - \dfrac{1024y^2}{36864} = \dfrac{36864}{36864}$

$$\frac{x^2}{64} - \frac{y^2}{36} = 1$$

$$\frac{x^2}{8^2} - \frac{y^2}{6^2} = 1$$

Asymptotes are $y = \frac{b}{a}x$ and $y = \frac{^-b}{a}x$

$$y = \frac{6}{8}x \quad \text{and} \quad y = \frac{^-6}{8}x$$

$$y = \frac{3}{4}x \quad \text{and} \quad y = \frac{^-3}{4}x$$

Asymptotes are $y = \pm \frac{3}{4}x$

(37) $\qquad \frac{y^2}{144} - \frac{x^2}{81} = 1$ is in standard form

center at $(0, 0)$

When $x = 0$

$$\frac{y^2}{144} - \frac{0^2}{81} = 1$$

$$\frac{y^2}{144} = 1$$

$$y^2 = 144$$

$$y^2 = 12^2$$

$$y = 12 \quad \text{or} \quad y = ^-12$$

y-intercepts (vertices) = $(0, 12)$ and $(0, ^-12)$

when $y = 0$

$$\frac{0^2}{144} - \frac{x^2}{81} = 1$$

$$-\frac{x^2}{81} = 1$$

$$x^2 = ^-81 \quad \text{No real solution}$$

$$\frac{y^2}{144} - \frac{x^2}{81} = 1$$

$$\frac{y^2}{12^2} - \frac{x^2}{9^2} = 1,$$

Compare with $\dfrac{y^2}{a^2} - \dfrac{x^2}{b^2} = 1$

$$a = 12 \text{ and } b = 9$$

Asymptotes are $y = \dfrac{a}{b}x$ and $y = \dfrac{^-a}{b}x$

$$y = \frac{12}{9}x \text{ and } y = \frac{^-12}{9}x$$

$$y = \frac{4}{3}x \text{ and } y = \frac{^-4}{3}x$$

Asymptotes are $y = \pm\dfrac{4}{3}x$

(38) $y^2 - x^2 - 4y - 6x - 41 = 0$ by completing squares

$y^2 - 4y - x^2 - 6x - 41 = 0$

$y^2 - 4y + \left(\frac{4}{2}\right)^2 - x^2 - 6x + \left(\frac{6}{2}\right)^2 - 41 = 0 + \left(\frac{4}{2}\right)^2 + \left(\frac{6}{2}\right)^2$

$y^2 - 4y + 2^2 - x^2 - 6x + 3^2 - 41 = 0 + 2^2 + 3^2$

$y^2 - 4y + 4 - x^2 - 6x + 9 - 41 = 0 + 4 + 9$

$y^2 - 4y + 4 - x^2 - 6x + 9 = 54$

$y^2 - 4y + 4 - x^2 - 6x + 9 - 18 = 54 - 18$

$y^2 - 4y + 4 - x^2 - 6x - 9 = 36$

$y^2 - 4y + 4 - (x^2 + 6x + 9) = 36$

$(y - 2)^2 - (x + 3)^2 = 36$

$$\frac{(y - 2)^2}{36} - \frac{(x + 3)^2}{36} = \frac{36}{36}$$

$$\frac{(y - 2)^2}{36} - \frac{(x + 3)^2}{36} = 1$$

$$\frac{(y-2)^2}{6^2} - \frac{(x+3)^2}{6^2} = 1$$

From $\quad \dfrac{(y-k)^2}{a^2} - \dfrac{(x-h)^2}{b^2} = 1$

$$k = 2, \qquad h = {}^-3, \qquad a = b = 6$$

When a = b, the hyperbola is called <u>equilateral hyperbola</u>

Equations of asymptotes are

$$y-k = \pm\frac{a}{b}(x-h)$$

$$y-2 = \pm\frac{6}{6}(x-{}^-3)$$
$$y-2 = \pm(x+3)$$

(39) $\qquad\qquad x^2 = 8(y+2)$

Rewrite in this form $(x-0)^2 = 8(y+2)$

Compare with $\quad (x-h)^2 = 4p(y-k)$

$$h = 0, \quad k = {}^-2$$

vertex $(h, k) = (0, {}^-2)$

$$4p = 8,$$
$$p = 2$$

p is positive opens upward

For intercept(s)   Let $y = 0$

$$x^2 = 8(y+2)$$
$$x^2 = 8(0+2)$$
$$x^2 = 8(2)$$
$$x^2 = 16$$

introduce square roots on both sides

$$\sqrt{x^2} = \sqrt{16}$$
$$\sqrt{x^2} = \sqrt{4^2}$$
$$x = \pm4$$

or $\quad x = 4, \; x = {}^-4$

$\therefore$ x-intercepts are $(4, 0)$ and $({}^-4, 0)$

When $x = 0$

$$x^2 = 8(y + 2)$$
$$0^2 = 8(y + 2)$$
$$0 = 8y + 16$$

Subtract $8y$ on both sides

$$^-8y = 16$$

Divide by $^-8$ on both sides

$$\frac{^-8y}{^-8} = \frac{\overset{^-2}{16}}{^-8}$$

$$y = ^-2$$

$\therefore$ y-intercept is $(0, ^-2)$

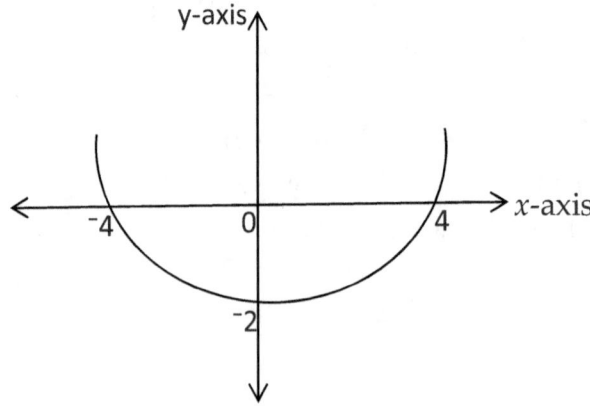

(40)
$$PF_1 + PF_2 = 10$$
$$\sqrt{(x - 0)^2 + (y - 4)^2} + \sqrt{(x - 0)^2 + (y + 4)^2} = 10$$
$$\sqrt{x^2 + y^2 - 8y + 16} + \sqrt{x^2 + y^2 + 8y + 16} = 10$$
$$\sqrt{x^2 + y^2 - 8y + 16} = 10 - \sqrt{x^2 + y^2 + 8y + 16}$$

Square both sides

$$x^2 + y^2 - 8y + 16 = 100 - 20\sqrt{x^2 + y^2 + 8y + 16} + x^2 + y^2 + 8y + 16$$

$$^-8y = 100 - 20\sqrt{x^2 + y^2 + 8y + 16} + 8y$$
$$20\sqrt{x^2 + y^2 + 8y + 16} = 8y + 8y + 100$$
$$20\sqrt{x^2 + y^2 + 8y + 16} = 16y + 100 \quad \text{square both sides}$$

$$400(x^2 + y^2 + 8y + 16) = 256y^2 + 3200y + 10000$$
$$400x^2 + 400y^2 + 3200y + 6400 = 256y^2 + 3200y + 10000$$
$$400x^2 + 400y^2 - 256y^2 = 10000 - 6400$$
$$400x^2 + 144y^2 = 3600$$
$$\frac{400x^2}{3600} + \frac{144y^2}{3600} = \frac{3600}{3600}$$

$$\frac{x^2}{9} + \frac{y^2}{25} = 1$$
$$c = 4$$
$$2a = 10,$$
$$a = 5$$

Center $(0, 0)$

Vertex $(a, 0)$ and $(^-a, 0) = (5, 0)$ and $(^-5, 0)$

OR $\qquad c = 4$
$$2a = 10,$$
$$a = 5$$
$$c^2 = a^2 - b^2$$
$$4^2 = 5^2 - b^2$$
$$16 = 25 - b^2$$
$$b^2 = 25 - 16$$
$$b^2 = 9$$
$$b = 3$$

From $\quad \dfrac{x^2}{b^2} + \dfrac{y^2}{a^2} = 1$

$$\frac{x^2}{3^2} + \frac{y^2}{5^2} = 1$$

$$\frac{x^2}{9} + \frac{y^2}{25} = 1$$

Center $(0, 0)$

Vertex $(a, 0)$ and $(^-a, 0) = (5, 0)$ and $(^-5, 0)$

(41)

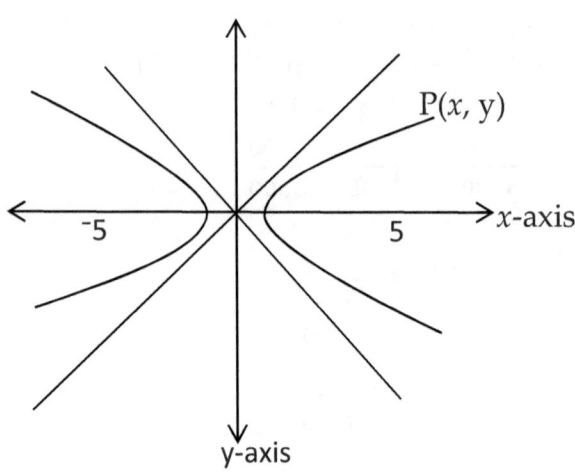

center $(0, 0)$

F_1 and F_2 at $(0, 5)$ and $(0, {}^-5)$

$c = 5$

From $|PF_1 - PF_2| = 8$

$2a = 8$

$a = \dfrac{8}{2}$

$a = 4$

From $c^2 = a^2 + b^2$

$5^2 = 4^2 + b^2$

$b^2 = 5^2 - 4^2$

$b^2 = 25 - 16$

$b^2 = 9$

$b = 3$

From $\dfrac{y^2}{a^2} - \dfrac{x^2}{b^2} = 1$

$\dfrac{y^2}{4^2} - \dfrac{x^2}{3^2} = 1$

$\dfrac{y^2}{16} - \dfrac{x^2}{9} = 1$

Asymptotes are $y = \dfrac{a}{b}x$ and $y = \dfrac{{}^-a}{b}x$

$$y = \dfrac{4}{3}x \quad \text{and} \quad y = \dfrac{{}^-4}{3}x$$

Asymptotes are $y = \pm\,\dfrac{4}{3}x$

OR From $|PF_1 - PF_2| = 8$

$|\sqrt{(x-0)^2 + (y-5)^2} - \sqrt{(x+0)^2 + (y+5)^2}| = 8$

$\sqrt{x^2 + y^2 - 10y + 25} - \sqrt{x^2 + y^2 + 10y + 25} = \pm 8$

$\sqrt{x^2 + y^2 - 10y + 25} = \pm 8 + \sqrt{x^2 + y^2 + 10y + 25}$

Square both sides

$x^2 + y^2 - 10y + 25 = 64 \pm 16\sqrt{x^2 + y^2 + 10y + 25} + x^2$
$$+ y^2 + 10y + 25$$

$-10y = 64 \pm 16\sqrt{x^2 + y^2 + 10y + 25} + 10y$

$\pm 16\sqrt{x^2 + y^2 + 10y + 25} = 10y + 10y + 64$

$\pm 16\sqrt{x^2 + y^2 + 10y + 25} = 20y + 64$

Square both sides

$256(x^2 + y^2 + 10y + 25) = 400y^2 + 2560y + 4096$

$256x^2 + 256y^2 + 2560y + 6400 = 400y^2 + 2560y + 4096$

$256x^2 + 256y^2 - 400y^2 = 4096 - 6400$

$256x^2 - 144y^2 = {}^-2304$

$^-144y^2 + 256x^2 = {}^-2304$

Multiply both sides by $^-1$

$144y^2 - 256x^2 = 2304$

$$\dfrac{144y^2}{2304} - \dfrac{256x^2}{2304} = \dfrac{2304}{2304}$$

$$\dfrac{y^2}{16} - \dfrac{x^2}{9} = 1$$

$$\dfrac{y^2}{4^2} - \dfrac{x^2}{3^2} = 1$$

Asymptotes are $y = \dfrac{a}{b}x$ and $y = \dfrac{^-a}{b}x$

$y = \dfrac{4}{3}x$ and $y = \dfrac{^-4}{3}x$

Asymptotes are $y = \pm\dfrac{4}{3}x$

(42) $4x^2 - 9y^2 - 24x - 36y - 36 = 0$

By completing squares

$4x^2 - 24x - 9y^2 - 36y - 36 = 0$

$4x^2 - 24x + \left(\dfrac{24}{2}\right)^2 - 9y^2 - 36y + \left(\dfrac{36}{2}\right)^2 - 36 = 0 + \left(\dfrac{24}{2}\right)^2 +$

$\left(\dfrac{36}{2}\right)^2$

$4x^2 - 24x + 12^2 - 9y^2 - 36y + 18^2 - 36 = 0 + 12^2 + 18^2$

$4x^2 - 24x + 144 - 9y^2 - 36y + 324 - 36 = 0 + 144 + 324$

$4x^2 - 24x + 144 - 9y^2 - 36y + 288 = 468$

$4x^2 - 24x - 9y^2 - 36y + 432 - 432 = 468 - 432$

$4x^2 - 24x - 9y^2 - 36y = 36$

$4x^2 - 24x + 36 - 36 - 9y^2 - 36y = 36$

$4x^2 - 24x + 36 - 9y^2 - 36y - 36 = 36$

$4x^2 - 24x + 36 - (9y^2 + 36y + 36) = 36$

$4(x^2 - 6x + 9) - 9(y^2 + 4y + 4) = 36$

$4(x - 3)^2 - 9(y + 2)^2 = 36$

$\dfrac{4(x-3)^2}{36} - \dfrac{9(y+2)^2}{36} = \dfrac{36}{36}$

$\dfrac{(x-3)^2}{9} - \dfrac{(y+2)^2}{4} = 1$

$\dfrac{(x-3)^2}{3^2} - \dfrac{(y+2)^2}{2^2} = 1$ compare with

$\dfrac{(x-h)^2}{a^2} - \dfrac{(y-k)^2}{b^2} = 1$

$$h = 3 \qquad k = {}^-2 \qquad a = 3 \qquad b = 2$$

Equations of asymptotes are

$$y - k = \pm \frac{b}{a}(x - h)$$

$$y - {}^-2 = \pm \frac{2}{3}(x - 3)$$

$$y + 2 = \pm \frac{2}{3}(x - 3)$$

Exercise 15B

(1) What is the name of 10 sided polygon?

(2) What kind of angles are X, Y and Z?

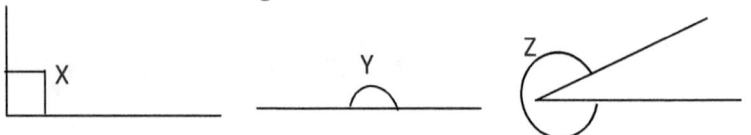

(3) Is a regular polygon an equilateral polygon, an equiangular polygon or it is both equilateral and equiangular polygon?

(4)

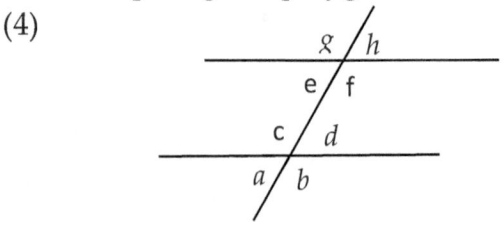

(a) Given $a° = 75°$; find all other angles and their relationship between each other.

(b) Which angles are congruent angles?

5(a) Find an interior angle of a regular hexagon

(b) Find an interior angle of a regular octagon

(c) Find an exterior angle of a regular nonagon

(6) The length of a regular decagon is 7m and its apothem is 9m. Find the area of a regular decagon.

(7) The side of a regular pentagon is 16m and the radius is 10m. find the area of a regular pentagon.

(8) The side of a regular pentagon is 24m and the radius is 15m. find the area of a regular pentagon

(9) Find the perimeter of an equilateral triangle with side measure 21.2in.

(10) The vertex angle of an isosceles triangle is 91.2° Find the base angle.

(11) Find the perimeter and the area of a triangle bellow.

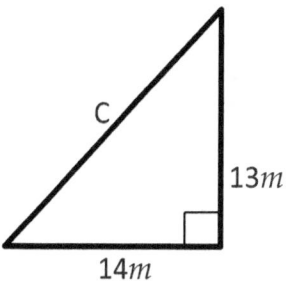

(12) The triangles area similar and the sides are proportion Find *a*.

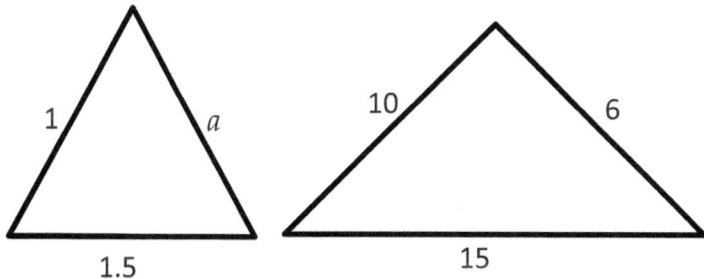

(13) Two triangles said to be Congruent. Find angle *x*

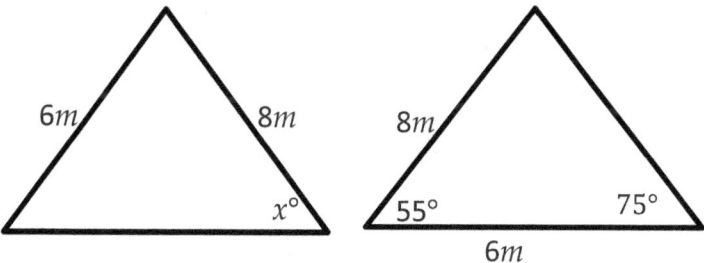

(14) The perimeter of a rectangle is *54m,* its length is *15m* and its width is *4x*. Find *x*

(15) The length of a rectangle is three times the width. Find the width given the area of a rectangle is *243m²*.

(16) The length of a rectangle is nineteen more than the width. Find width given the perimeter of the rectangle is *98m*.

(17) Find the area of the following figures:
Given *a = 9m, b = 16m, h = 8m* and diameter *D = 12m*

(a)

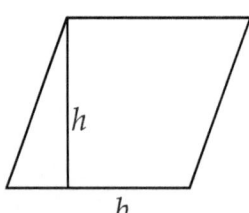

(b)

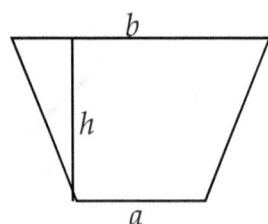

(c)

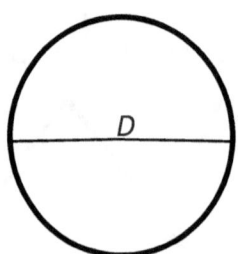

(18) Find the volume of figures given $w = 2.5m$, $L = 12.8m$, $h = 5m$ and for cylinder and cone diameter $D = 10m$.

(a)

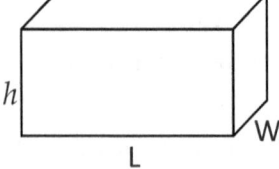

(b)

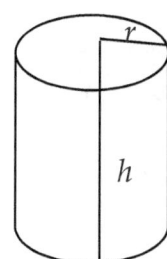

(c)

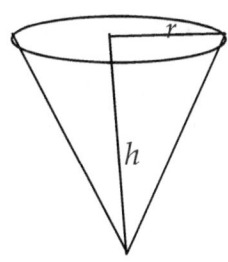

(19)   The following figures are filled with water.
Given the diameter of the cylinder and cone $D = 8m$,
$h = 12m$, $h_1 = 7m$, $h_2 = 7m$ and $r_2 = 3m$. Find the
volume of the empty space that is not filled with water.

(a)

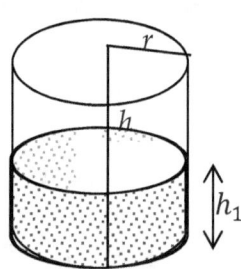

(b)

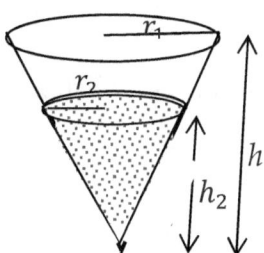

(20) The vertex angle of an isosceles triangle exceeds the measurement of each base angle by $60°$. Find the value of vertex angle of the triangle.

(21) Find a perimeter of a triangle below

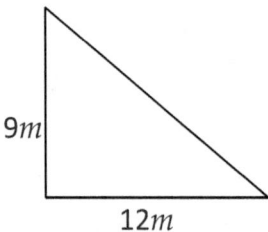

9m

12m

(22) Find the arc length AB with radius $BC = 16m$

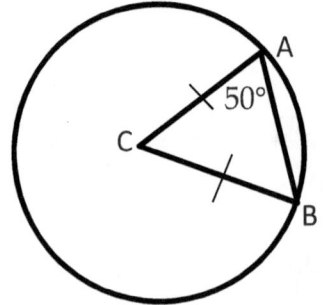

A

50°

C

B

(23) Find the equation of the circle with center at origin and radius 14.

(24) Find the center and radius of the equation of a circle $x^2 + y^2 = 144$.

(25) Find equation of the circle with center at $(4, ^-7)$ and with radius 8.

(26) Find the center and the radius of the circle with equation $2x^2 + 2y^2 + 4x - 20y + 9 = 29$.

(27) Find the equation of a parabola with vertex at origin and focus $(8, 0)$.

(28) Find the equation of a parabola with vertex at origin and focus $(0, 11)$.

(29) Find the vertex, focus and diretrix of the equation of a Parabola $y^2 = ^-68x$.

(30) Find the vertex, focus and diretrix of the equation of a parabola $x^2 = -52y$.

(31) Find the equation of a parabola opening right with vertex $(9, 2)$ and the focus $(12, 4)$.

(32) Find the vertex, focus and diretrix of the equation of a parabola $(x - 11)^2 = 36(y-5)$

(33) Sketch and find the center and the ends of the axes of ellipse $\dfrac{x^2}{49} + \dfrac{y^2}{16} = 1.$

(34) Find equation of an ellipse, its center and vertex given the foci $(12, 0)$ and $(-12, 0)$, the sum of focal radii is 30.

(35) A graph of $3x^2 + 4y^2 + 36x - 16y + 112 = 0$ is an ellipse with center $(-6, 2)$ and axes along $x = -6$ and $y = 2$ put it into a standard form.

(36) Find the equation of hyperbola and its asymptotes with center at $(0, 0)$, foci F_1 and F_2 at $(15, 0)$ and $(-15, 0)$ and also with $|PF_1 - PF_2| = 24$

(37) Find the center, vertices and asymptotes of the hyperbola $\dfrac{y^2}{64} - \dfrac{x^2}{36} = 1.$

(38) Write the equation $y^2 - x^2 - 10y - 4x + 5 = 0$ in standard form of hyperbola and find its equation of asymptotes.

(39) Sketch the graph of $x^2 = 9(y + 4)$ and find the vertex and the intercept(s).

(40) Find the equation of an ellipse, its center and the vertex given the foci $(0, 8)$ and $(0, -8)$, the sum of focal radii is 20

(41) Find the equation of hyperbola and its asymptotes with center at $(0, 0)$, foci F_1 and F_2 at $(0, 10)$ and $(0, -10)$ and also with $|PF_1 - PF_2| = 16$

(42) write the equation $9x^2 - 16y^2 - 72x - 96y - 144 = 0$ in standard form of hyperbola and find its equation of asymptotes.

DAVID M KASASA

ANSWERS FOR EXERCISE 1B

(1) $^+133$ (2) $P = \{61, 67, 71, 73, 79, 83, 89\}$

(3) Numbers < 9 and > 0

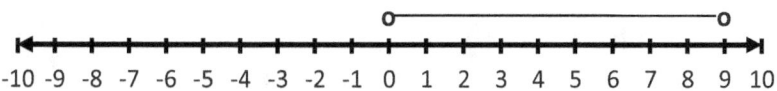

(4) $19 + {}^-15$
    $19 - 15 = {}^+4$

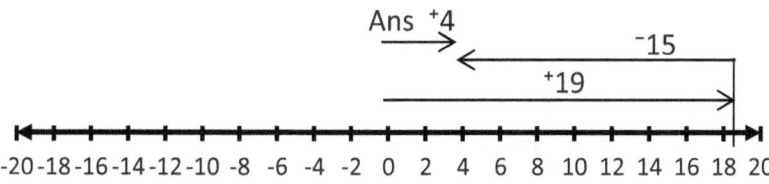

(5) $^-8 - {}^-16 + 2$
    $^-8 + 16 + 2 = {}^+10$

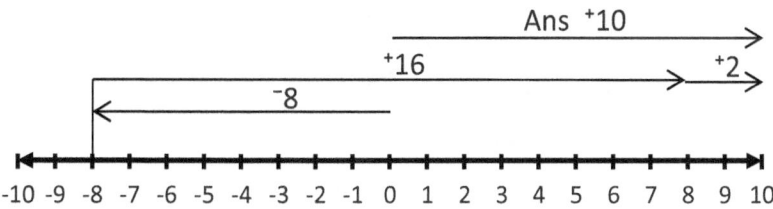

(6) $11 + ({}^-17) + 3$
    $11 - 17 + 3 = {}^-3$

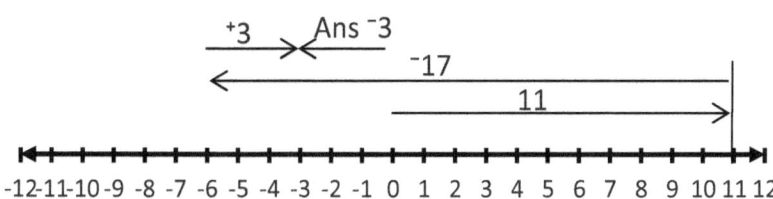

(7) ⁻55   (8) 3   (9) 0   (10) 60

(11) 0   (12) ⁻13   (13) 72   (14) ⁻26

(15) $I = \{......, ⁻30, ⁻25, ⁻20, ⁻15, \underline{⁻10}, \underline{⁻5}, \underline{0} ...\}$

  $N = \{1, 2, 4, 8, \underline{16}, \underline{32}, \underline{64}, ...\}$

  $V = \{-1, -3, -6, -10, \underline{⁻15}, \underline{⁻21}, \underline{⁻28}, ...\}$

  $W = \{0, 5, 10, 15, \underline{20}, \underline{25}, \underline{30}, ...\}$

  $P = \{......, 5, 7, 11, 13, \underline{17}, \underline{19}, \underline{23}, ...\}$

  $E = \{..., ⁻8, ⁻4, 0, 4, \underline{8}, \underline{12}, \underline{16}....\}$

  $O = \{...⁻7, ⁻3, 1, 5, \underline{9}, \underline{13}, \underline{17}, ...\}$

(16) $x_4 = 201$

17(a) $U = \{8, 9, 10, 11, 12, 13, 14, 15, 16, 17\}$

b(i) false   (ii) false   (iii) true   (iv) false

 (v) true   (vi) false

c(i) 14   (ii) 14   (iii) 14

(iv) $A \cup B = \{8, 9, 10, 11, 12, 13, 14\}$

(v) $A \cup B \cup C = U = \{8, 9, 10, 11, 12, 13, 14, 15, 16, 17\}$

(vi) $A' \cup B = \{8, 9, 11\}$

(d)

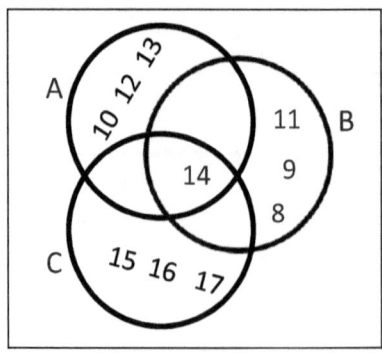

(18)   300 feet

(19) $N \cap W \cup P' = \{1, 4, 6, 8, 9\}$

U = Reals

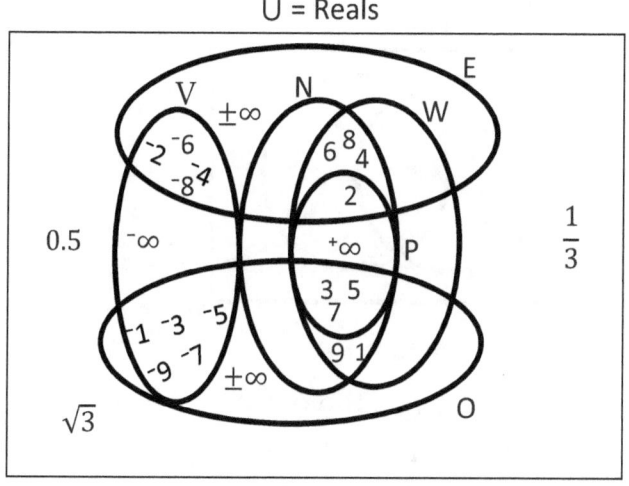

(20) 9

21(i) 40   (ii) 35   (iii) 10   (iv) 5

(22) $P = \{23, 29, 31\}$

   $O = \{21, 23, 25, 27, 29, 31, 33, 35\}$

   $E = \{20, 22, 24, 26, 28, 30, 32, 34\}$

(i) O∩P = {23, 29, 31}

(ii) P∩E = ∅

(iii) O∩P∩E = ∅

(iv) O∪P∪E = {20, 21, 22, 23, 24, 25, 26, 27,
                  28, 29, 30, 31, 32, 33, 34, 35}

(v) O∩E∪P' = ∅

(iv) Venn diagram

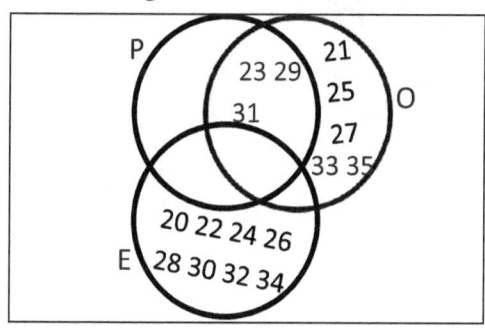

(23) The number of people who like sports is *12*
Venn diagram

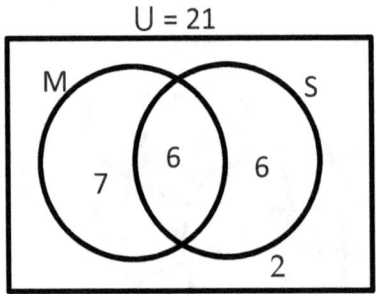

24(i) Math and Science only $M \cap S \cup E' = 7$
(ii) Math and English only $M \cap E \cup S' = 2$
(iii) Science and English only $S \cap E \cup M' = 4$
(iv) Science only S only $= 12$
(v) Math only M only $= 10$
(vi) English only E only $= 15$
(vii) Not any of the three $S' \cup M' \cup E' = 5$

Venn diagram

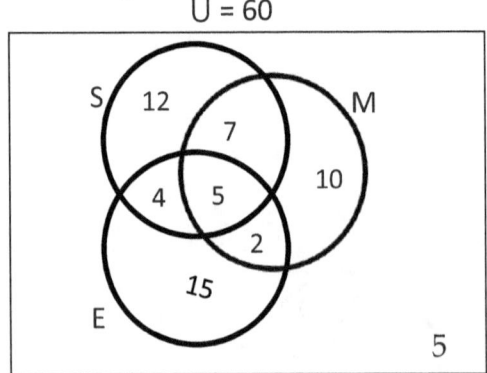

ANSWERS FOR EXERCISE 2B

(1) The bigger fraction is $\dfrac{^-7}{8}$, their sum $= \dfrac{^-49}{24}$

(2) $\dfrac{17}{44}$ (3) $\dfrac{97}{60}$ (4) $2\dfrac{2}{15}$ (5) $^-5\dfrac{5}{6}$ (6) $\dfrac{2}{3}$

(7) $5\dfrac{1}{2}$ (8) 9 (9) $\dfrac{2}{3}$ (10) $\dfrac{1}{5}$ (11) $\dfrac{3}{2}$

(12) 1 (13) $\dfrac{^-39}{40}$ (14) $^-5$ (15) 6 (16) 4

(17) $\dfrac{^-5}{6}$ (18) $\dfrac{x+y}{4xy}$ (19) $\dfrac{ab(c+d)}{cd}$ (20) c

ANSWERS FOR EXERCISE 3B

(1) 1 (2) 100 (3) 256 (4) 512 (5) $9(xy)^2$

(6) $\left(\dfrac{y}{x}\right)^2$ (7) $2p^2q^3$ (8) $\dfrac{4}{9}$ (9) 3 (10) 36

(11) 3 (12) 125 (13) $12p^2q$ (14) 8 (15) $2^{\frac{5}{3}}$
(16) 1.213×10^3 (17) 5.76×10^{-1}
(18) 8.2×10^{-4} (19) 5.4318×10^2
(20) 9.999×10^4 (21) 1.0×10^1

ANSWERS FOR EXERCISE 4B

1(a) 0.375 It is Terminating Decimal
(b) 0.889 It is Repeating Decimal

(c) 0.692   It is Irrational number

2(i) 0.316 is bigger   (ii) 0.316 ≈ 0.32, 0.0316 ≈ 0.03

(3) ⁻0.0097 is bigger   4(a) 1   (b) 40.2   (c) 16.2

(d) 11.98   5(a) 31.584   (b) 3.24   (c) 9   (d) 0.47

(e) 900   6(i) 0.36   (ii) 0   (iii) 0.2

ANSWERS FOR EXERCISE 5B

1(a) $\dfrac{24}{25}$ (b) $\dfrac{15}{16}$ 2(a) 260%   (b) 45.6%   (c) 10000%

3(a) $\dfrac{7}{20}$ (b) $\dfrac{1}{11}$ (c) $\dfrac{1}{6}$ (d) $\dfrac{11}{2}$ (e) $\dfrac{3}{400}$

4(a) 2.5   (b) 0.00195   (c) 0.005   (d) 0.3333

(e) 1.65   5(a) 150   (b) 0.24   (c) 1   6(a) 5%

(b) 0.1%   (c) 2%   (7) 24girls   (8) 220girls

(9) 60%

ANSWERS FOR EXERCISE 6B

(1) 12.3   (2) $11\sqrt{3}$ (3) $9\sqrt{5} + 4\sqrt{3}$ (4) $2(2\sqrt{2} + \sqrt{3})$

(5) $^{-}\sqrt{7a}$ (6) $2b\sqrt[3]{3}$ (7) $3y\sqrt[3]{2y}$ (8) 7

(9) 32   (10) $\dfrac{1}{25}$ (11) $\sqrt[3]{xxyz^2}$ (12) $x\sqrt{3}$

(13) $\dfrac{3ab}{c}\sqrt{2b}$ (14) $4(xy)^4$ (15) $3a^3c^2\sqrt[4]{b^2}$ (16) 360

(17) $4\sqrt{2}$ (18) $\dfrac{3}{4}$ (19) $\dfrac{6x}{y}$ (20) $15 + 8\sqrt{6}$

(21) $\dfrac{-\sqrt[3]{3}}{2}$ (22) $5 - \sqrt{15}$ (23) 1   (24) $1 + \sqrt{-1}$

ANSWERS FOR EXERCISE 7B

(1) $^-4$ (2) $^-4$ (3) $^-16$ (4) $\dfrac{2}{3}$ (5) 11

(6) $\dfrac{1}{4}$ (7) 1   (8) 100   (9) 4   (10) 8

(11) 4   (12) 17   (13) 2   (14) $x > 6$

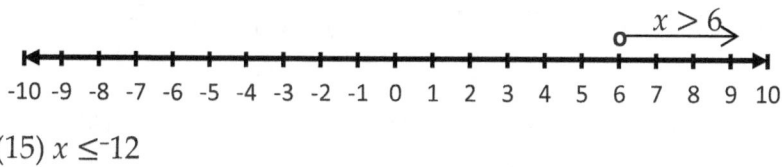

(15) $x \leq {}^-12$

(16) $x < {}^-6$ (17) $x \geq 2$ (18) $x > 102$ (19) $x \geq 17$

(20) $x = 0$ or $x = 8$

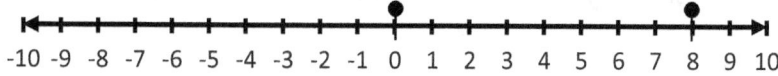

(21) $x < 4$ or $x > {}^-10$

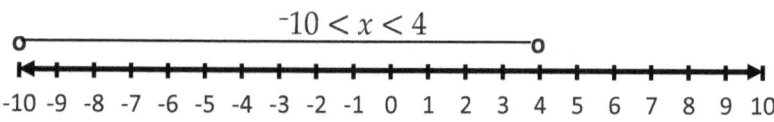

(22) $x \geq 8$ or $x \leq {}^-4$ (23) $^-14 \leq x \leq 18$

(24) $b = \dfrac{2A}{h} - a$ (25) 19   (26) 200   (27) 38

(28) 21 and 23   (29) 50, 51 and 52   (30) 15 and 45

(31) 38   (32) 105   (33) ⁻12   (34) 10 one-dollar bills and 12 five-dollar bills   (35) C   (36) B

ANSWERS FOR EXERCISE 8B

1(i) $21x$ (ii) $x^2(x - y)$ (iii) $(x - 2)(x + 2)$

(2) $5a(3b^2 + c)$ (3) $6(ab)^2$ (4) $(2x + 1) + \dfrac{2}{x + 3}$

(5) $4xy(3x+y)$ (6) $2x^2 - 5xy - 3y^2$ (7) $xy[(xz+yw)(xz-yw)]$

(8) $5(x + 3)(x - 3)$ (9) $\dfrac{1}{9}x^2 + 4x + 36$ (10) $121 - 22x + x^2$

(11) $7x^2y$ (12) $4 - x$ (13) $\dfrac{4}{x-5}$ (14) $2(x - 6)$

(15) $x(2x - 3)$ (16) $x^3 + 3x - 2 + \dfrac{5}{x + 2}$ (17) 13

(18) 14   (19) 30   (20) 99   (21) $^-4x + 3y$

(22) $2x(2x - 1)$ (23) $8(2x + 3)(4x^2 - 6x + 9)$

(24) $\dfrac{1}{2}(x - 2)(x^2 + 2x + 4)$ (25) $(7x + 2)(7x - 2)$

(26) $\dfrac{x^2}{4} + 4y^2 + z^2 + 2xy + xz + 4yz$ (27) $\dfrac{x + 2y}{x + y}$

(28) $^-7$ (29) 9   (30) $\dfrac{2}{7}$ (31) $\dfrac{2}{x - 3} - \dfrac{1}{x + 3}$

(32) $\dfrac{2}{x + 4} + \dfrac{1}{x - 4}$

ANSWERS FOR EXERCISE 9B

(1) $(a + 2b)(3a + 4b)$ (2) no factors, it is a Prime Polynomial

(3) $(x + 1)(^-x + 10)$ (4) $\dfrac{2x - 1}{2x + 1}$ 5(i) $\dfrac{x + 3}{x + 2}$ (ii) $x = \{^-2, 4\}$

(iii) $x = \{^-3, 4\}$ (6) $x = \{^-2, ^-6\}$ (7) $x = \{\dfrac{^-1}{3}, ^-4\}$

(8) $x = 0$ (9) $x = \{1, \dfrac{1}{4}\}$ (10) $x = ^-3$ (11) $x = \{^-9, 0, 3\}$

(12) $x = \{^-1, 7\}$ (13) $x = \{^-3, 1\}$ (14) $x = \left\{\dfrac{^-6}{7}, \dfrac{6}{7}, 4\right\}$

(15) $x = \{5, 8\}$ (16) $x = \{8, 9\}$ (17) $x = ^-2 \pm \sqrt{7}$

(18) $x = 2 \pm i$ (19) $x = \{\pm 2, \pm\sqrt{5}\}$ (20) $x = 1 \pm i$

(21) 9 and 11

ANSWERS FOR EXERCISE 10B

1(a) $-p = \begin{bmatrix} ^-6 & ^-7 \\ ^-4 & 3 \end{bmatrix}$ $-q = \begin{bmatrix} 3 & ^-13 \\ ^-8 & 4 \end{bmatrix}$ (b) $\begin{bmatrix} 3 & 20 \\ 12 & ^-7 \end{bmatrix}$

(c) $\begin{bmatrix} 15 & 1 \\ 0 & ^-2 \end{bmatrix}$ (2) 29   (3) $\begin{bmatrix} 2 & 8 \\ 3 & 12 \end{bmatrix}$ (4) $\begin{bmatrix} 17 & 19 \\ ^-1 & ^-2 \end{bmatrix}$

(5) $\begin{bmatrix} ^-7 & ^-3 & 2 \\ 4 & 12 & 16 \end{bmatrix}$ (6) $\begin{bmatrix} 8 & 11 \\ \dfrac{1}{2} & 0 \end{bmatrix}$ 7(i) $a = 36$

$b = 3$ $c = 15$ $d = 9$ (ii) $a = 5$ $b = 4$

$c = 8$ $d = 20$ 8(a) 20   (b) $\begin{bmatrix} \dfrac{1}{10} & -\dfrac{3}{20} \\ \dfrac{1}{10} & \dfrac{7}{20} \end{bmatrix}$

(c) $P * P^{-1} = \begin{bmatrix} 1 & 0 \\ 0 & 1 \end{bmatrix}$ 9(a) 2   (b) $\begin{bmatrix} 1 & ^-3 \\ -\dfrac{1}{2} & 2 \end{bmatrix}$

(c) $q * q^{-1} = \begin{bmatrix} 1 & 0 \\ 0 & 1 \end{bmatrix}$

ANSWERS FOR EXERCISE 11B

(1) $x = 1$ $y = 2$ (2) $x = ^-2$ $y = 6$ (3) $x = ^-1$ $y = 1$

(4) $x = \dfrac{1}{3}$ $y = 2$ (5) $x = ^-\dfrac{1}{2}$ $y = 0$ (6) $x = 7$ $y = 3$

(7) $x = 5$ $y = ^-5$ (8) $x = 11$ $y = 12$ (9) $x = 5$ $y = ^-5$

(10) 16   (11) 10

ANSWERS FOR EXERCISE 12B

(1) K in 4th quadrant, L in 2nd quadrant, M in 3rd quadrant and N in 1st quadrant

2(a) For A Domain$(x) = \{ 1, 3, 5 \}$
           Range (y) $= \{ 0, ^-2, ^-4 \}$
     For B Domain$(x) = \{ 7, 2, 1 \}$
           Range (y) $= \{ 5, 5, 3 \}$
     For C Domain$(x) = \{ ^-1, 0, ^-1 \}$
           Range (y) $= \{ 4, 5, 2 \}$
     For D Domain$(x) = \{ 6, 4 \}$
           Range (y) $= \{ 4, 4 \}$

(b) A is a Function because all its Domain or x-values are different from each other.

B is a Function because all its Domain or x-values are different from each other

C is not a Function because it has same Domain or x-values which are $^-1$.

D is a Function because all its Domain or x-values are different from each other.

(c)

For $D = (6, 4), (4, 4)$

x	y
6	4
4	4

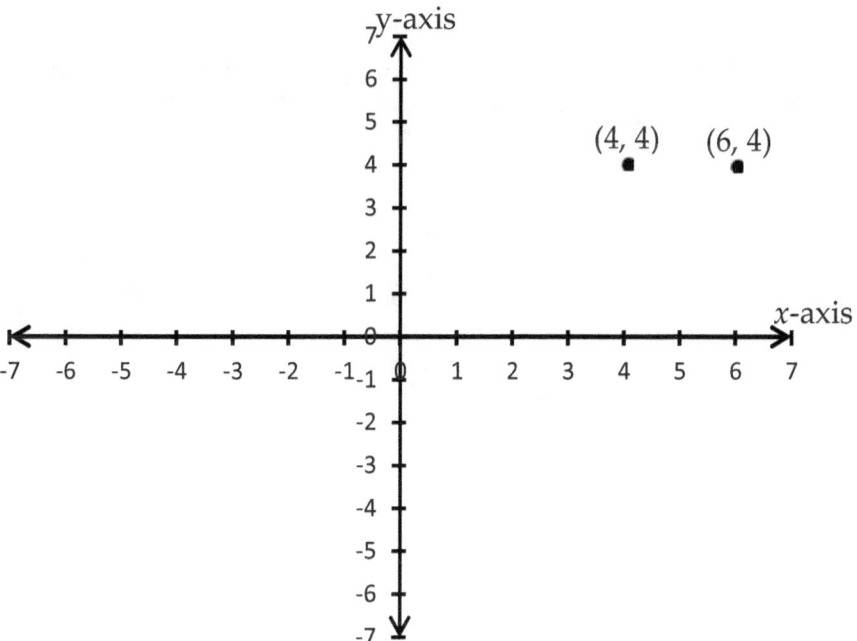

(3) $f_{(-1)} = {}^-6, \quad f_{(3)} = 2$ (4) Range $= \{\,{}^-1, 0, 1\,\}$ (5) $2a$

(6) $f_{(9)} = 6, \quad f_{(a+7)} = 3a$ 7(a) ${}^-1$ (b) $(x + 4)(x - 2)$

8(a) $\dfrac{x + 6}{11}$ (b) x (c) x 9(i) 9   (ii) 8   (iii) 25

(iv) 20 $\qquad$ 10(i) $f^{-1}_{(x)} = \dfrac{4}{3}x + \dfrac{8}{3}$ $\qquad$ (ii) $f^{-1}(f_{(x)}) = x$

(iii) f o $f^{-1}_{(x)} = x$

(11)

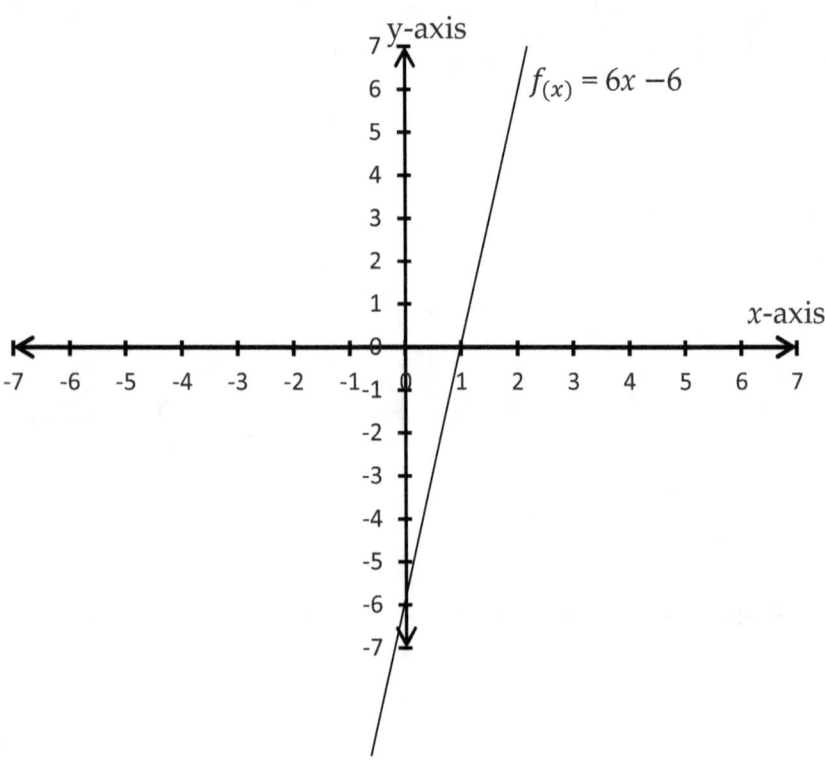

$f_{(x)} = 6x - 6$

(12) C

(13) $9x - y = 17$ $\qquad$ (14) $x + 3y = 24$ $\qquad$ (15) $y = 15x + 41$

16(i) 5 $\qquad$ (ii) $\left(\dfrac{-1}{2}, 2\right)$ $\qquad$ (iii) $\dfrac{4}{3}$ $\qquad$ (iv) $y = \dfrac{4}{3}x + \dfrac{8}{3}$ $\qquad$ (v) $\dfrac{4}{3}$

(vi) $\dfrac{-3}{4}$ $\qquad$ (vii) x-intercept is (-2, 0), y-intercept is $\left(0, \dfrac{8}{3}\right)$

(17) $\dfrac{3}{2}$ (18) $7x - y = 43$ 19(a) 60   (b) 8

(20) $2x - 3y = 13$ (21) $y = 2x - 11$ (22) $^{-}1$

(23) Perpendicular

(24)

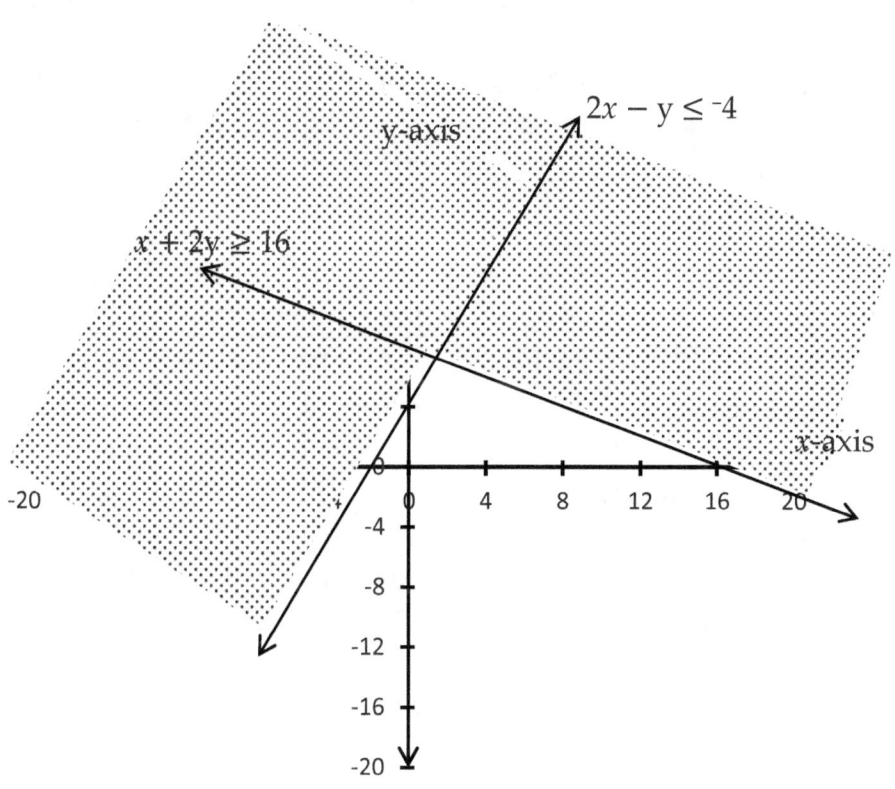

ANSWERS FOR EXERCISE 13B

(1) 81   (2) 125   (3) $\log_9 24$ (4) $\log_2 42$

(5) $\log_{10} 34$ (6) $\log_3 11$ (7) 10   (8) 13

(9) 8   (10) $^{-}6$ (11) $3\log_5 5p$ (12) 70

(13) $\dfrac{3}{4}$ (14) $\dfrac{\log_4 6}{3}$ (15) 20

ANSWERS FOR EXERCISE 14B

(1) $0.05km$ (2) $70m$ (3) 512 ounces   (4) 0.4 ton

(5) 1440 inches   (6) 240 yards   (7) 0.56 yard   (8) 50 pints

(9) 192 cups   (10) 4 tea spoons   (11) 0.01 gram

(12) 840 hours   (13) 10400 weeks   (14) 100°C   (15) 122°F

(16) $5500   (17) $7200   (18) $4420   (19) $264, $2464

(20) $9360   (21) $1866.67   (22) $90mi$

(23) $195mi$ (24) $\dfrac{4y}{3}$

ANSWERS FOR EXERCISE 15B

(1) Decagon   (2) X is a Right angle, Y is a straight angle and Z is a Reflex angle   (3) it is both equilateral and equiangular polygon.

4(a) $a° = d° = 75°$ Vertical angles

$b° = 180° - a°$ Straight angles(angles on line=180°)

$b° = 180° - 75°$

$b° = 105°$

$b° = c° = 105°$ Vertical angles

$d° = e° = 75°$ Alternate interior angles

$b° = g° = 105°$ Alternate exterior angles

$g° = f° = 105°$ Vertical angles

$h° = 180° - f°$ Straight angles(angles on line=180°)

$h° = 180° - 105°$

$h° = 75°$

(b) $a°$, $d°$, $e°$ and $h°$ are congruent angles also

$b°$, $c°$, $f°$ and $g°$ are congruent angles

5(a) 120°   (b) 135°   (c) 40°   (6) $315m^2$ (7) $240m^2$

(8) $540m^2$ (9) $63.6in$ (10) 44.4°   (11) $P = 46.1m$, $A = 91m^2$

(12) $a = 0.6$ (13) $50°$ (14) $x = 3m$ (15) $9m$ (16) $15m$

17(a) $128m^2$ (b) $100m^2$ (c) $113.04m^2$ 18(a) $160m^3$

(b) $392.5m^3$ (c) $130.83m^3$ 19(a) $251.2m^3$

(b) $135.02m^3$ (20) $100°$ (21) $36m$

(22) $22.33m$ (23) $x^2 + y^2 = 196$ (24) center at $(0, 0)$

   Radius $r = 12$ (25) $x^2 + y^2 - 8x + 14y + 1 = 0$

(26) center at $(^-1, 5)$, Radius $r = 6$ (27) $y^2 = 32x$

(28) $x^2 = 44y$ (29) vertex at $(0, 0)$, focus $(^-17, 0)$,

   the directrix is $x = 17$ (30) vertex $(0, 0)$, focus $(0, ^-13)$

   the directrix is $y = 13$ (31) $(y - 2)^2 = 12(x - 9)$

(32) vertex $(11, 5)$, focus $(11, 14)$, the directrix is $y = ^-4$

(33) x and y-intercepts $(7, 0)$, $(^-7, 0)$, $(0, 4)$ and $(0, ^-4)$

   Center $(0, 0)$

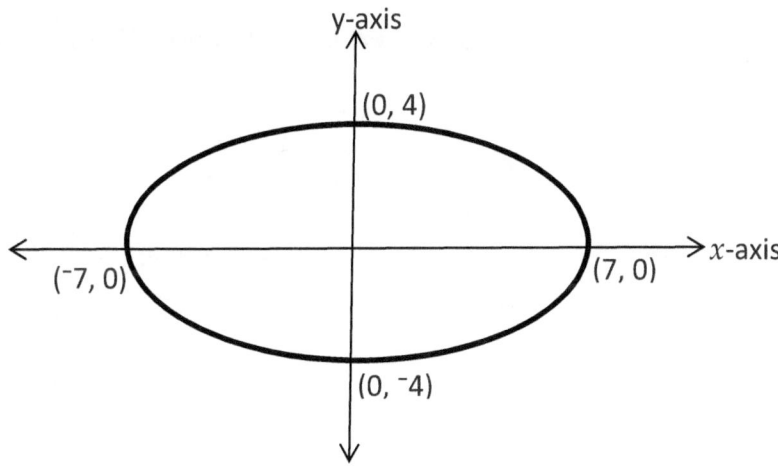

(34) $\dfrac{x^2}{225} + \dfrac{y^2}{81} = 1$, Center $(0, 0)$, Vertex $(15, 0)$ and $(^-15, 0)$

(35) $\dfrac{(x+6)^2}{4} + \dfrac{(y-2)^2}{3} = 1$ (36) $\dfrac{x^2}{144} - \dfrac{y^2}{81} = 1$, $y = \pm \dfrac{3}{4}x$

(37) Center $(0, 0)$, vertex $(0, 8)$ and $(0, ^-8)$, $y = \pm \dfrac{4}{3}x$

(38) $\dfrac{(y-5)^2}{16} - \dfrac{(x+2)^2}{16} = 1$, $y - 5 = \pm(x+2)$

(39) vertex(0, ⁻4), x and y-intercepts (6, 0), (⁻6, 0) and (0, ⁻4)

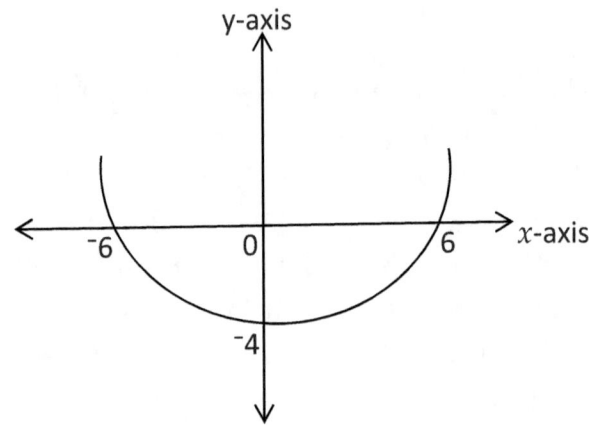

(40) $\dfrac{x^2}{36} + \dfrac{y^2}{100} = 1$, Center (0, 0), vertex (10, 0) and (⁻10, 0)

(41) $\dfrac{y^2}{64} - \dfrac{x^2}{36} = 1$, $y = \pm\dfrac{4}{3}x$ (42) $\dfrac{(x-4)^2}{16} - \dfrac{(y+3)^2}{9} = 1$,

$$y + 3 = \pm\dfrac{3}{4}(x - 4)$$

INDEX

ABOUT THE AUTHOR

David Kasasa lives in Boston Massachusetts. He has been teaching Mathematics for over *15* years and during this time he identified many challenges that students face which compelled him to write this simplified book on Algebra *I* and *II*. For more information visit www.davidkasasa.com